APQ LIBRARY OF
PHILOSOPHY

APQ LIBRARY OF
PHILOSOPHY

Edited by NICHOLAS RESCHER

INTRODUCTION
TO THE
PHILOSOPHY OF
MATHEMATICS

HUGH LEHMAN

ROWMAN AND LITTLEFIELD
TOTOWA, NEW JERSEY

First published in the United States, 1979 by
Rowman and Littlefield, Totowa, New Jersey

Library of Congress Cataloging in Publication Data

Lehman, Hugh.
Introduction to the philosophy of mathematics.

(APQ library of philosophy)
1. Mathematics—Philosophy. I. Title. II. Series.
QA8.4.L44   510'.1   78-9934
ISBN 0-8476-6109-1

Printed in Great Britain

TABLE OF CONTENTS

PART II: MATHEMATICAL KNOWLEDGE

CHAPTER FIVE: FICTIONALISM, PROOF, GÖDEL'S VIEW OF MATHEMATICAL KNOWLEDGE

PREFACE

It gives me great pleasure to thank the editor of this series, Nicholas Rescher, for his encouragement and assistance with the publication of this book, and *Canada Council* for assistance with funds for typing this manuscript and to Mrs. Judy Martin and Miss Sheila MacPherson for typing.

I also wish to acknowledge my indebtedness to my parents and teachers.

This book is dedicated to my wife and children.

Introduction

1. My aim in this book is to discuss certain ontological and epistemological issues. In particular, I assume that modern man is in possession of a good deal of mathematical knowledge and that such knowledge is extensively applied in his scientific, artistic and practical endeavors. Further, I assume that, at the very least, knowledge is true, completely justified belief. That is to say, that if a person knows some mathematical proposition such as "There is a real number n which is the sum of the real numbers p and q", then he believes this proposition, his belief is true and his belief is completely justified. (I am not saying that these three conditions are sufficient for a person to have knowledge.) The ontological and epistemological questions which arise given these assumptions about knowledge are fairly obvious. Given that mathematical propositions are known, it follows that some mathematical propositions are true. The ontological question that immediately comes to mind then is: Are there mathematical entities? The point of view of common sense is perhaps, that if a proposition is true, it is because there are entities existing independently of the proposition which have the properties or stand in the relations which the proposition asserts of them. This suggests that since mathematical propositions are true, that there are entities in virtue of which the propositions are true. The ontological issue is whether there are such entities and if so what their nature is. The epistemological question which comes to mind is how mathematical beliefs come to be completely justified. These are the ontological and epistemological questions with which we shall be concerned in this work.

2. Many works in the philosophy of mathematics presuppose, on the part of the reader, extensive knowledge of mathematical logic, set theory or of some other branches of mathematics. Since many people, even people who are quite interested in ontological and epistemological issues, feel themselves to be lacking in regard to the requisite mathematical knowledge, they tend to ignore philosophy of mathematics. This is an undesirable course to take. Mathematics bulks so large in our lives that it is incumbent upon epistemologists and ontologists to understand it. A theory of

knowledge which professes to be complete must take mathematical knowledge into account. An ontological theory which ignores problems relating to the existence of mathematical entities is also incomplete.

I have tried in this work to write in such a way as to be understandable even by those who have not studied logic and mathematics extensively. There is danger in writing in this way of ignoring parts of mathematics that are not normally learned in elementary school. In order to avoid this danger I have introduced in one of two places some rather elementary parts of the theory of real numbers. Statements from this theory are typical of many areas of mathematical knowledge. Further, in discussing views of other philosophers, discussion of other mathematical examples has been necessary. To make such discussion intelligible to non-mathematical readers I have tried to explain mathematical terminology as briefly as possible. (The reader of a work in philosophy may not be interested in lengthy explanations of mathematical concepts.)

3.   My views concerning the ontological issues which are discussed in this work are (1) that there are mathematical entities and (2) that justification of mathematical beliefs requires sensory observations. Another way of expressing my view is to say that I accept a "realistic" interpretation of mathematical propositions and an empirical view of mathematical knowledge. The discussion which follows in the following chapters is intended to support these views.

My approach to supporting these views will be, in a sense, indirect. I shall first investigate a range of theories which hold that even though mathematical propositions are true or useful in science and everyday life, that there are no mathematical entities. Such theories fall into two groups. In one group are theories which hold that while mathematical statements are true, they do not imply the existence of mathematical entities. In the other group are theories which hold that mathematical statements are extremely useful and they do imply the existence of mathematical entities but that all such mathematical statements are false. According to theories in this second group, since mathematical statements are false, the entities, which are asserted or implied to exist by the statements of such theories, do not exist. I shall criticize views of both sorts and thereby hope to establish that since some mathematical statements are true, there are some mathematical entities.

Secondly I shall investigate a range of theories of mathematical knowledge which agree in holding that sensory observation does not enter into the complete justification of mathematical beliefs and shall try to show that such theories are unsatisfactory. Thirdly, I shall investigate some theories of knowledge which hold that our knowledge of mathematical propositions does rest on sensory observations. While some theories in this category are unsatisfactory, I shall try to show that it is a theory in this category which can give a satisfactory account of our mathematical knowledge.

4. Mathematical theories, at least ostensibly, carry ontological implications. That is, statements of the theories assert or imply that there exist mathematical entities. This is perfectly clear, for example, in the theory of real numbers. If one studies calculus in secondary school or university one will learn some theorems affirming the existence of limits of sequences, derivatives, definite integrals, etc. The proofs of these theorems depend on the axioms of the theory of real numbers. While these axioms may be formulated in various ways, one way of expressing them is as follows: There exists a non-empty set of objects, called the real numbers, which satisfies the following properties:

1. The set is closed under operations of addition and multiplication. (That is, whenever one adds or multiplies real numbers the sum or product is a real number.)
2. Addition and multiplication are associative and commutative. (That is, for any real numbers a, b, and c, $(a+b)+c = a+(b+c)$ and $a+b = b+a$. Similarly, $a \times (b \times c) = (a \times b) \times c$ and $(a \times b) = (b \times a)$.)
3. There exists in the set both additive and multiplicative identities. (That is for any real number n there is a real number r such that $n+r = n$, and for any real number n there is a real number r such that $n \times r = n$ ["$n \times r$" is read "n times r"]. The additive identity, of course, is normally called 'zero' in English and is written with the symbol '0'. The multiplicative identity is called "one" in English and is written with the symbol '1'.)
4. For any real number n, there exists a number m such that $n+m = 0$ (m is called the negative of n).
5. For any number n other than 0, there is a number m, such that $n \times m = 1$ (m is called the reciprocal of n).

6. For any real numbers n, m and q, we have that $n \times (m+q) = (n \times m) + (n \times q)$.

Objects which satisfy the above six axioms are referred to as a field. Thus, the axioms which are assumed in proofs of theorems of calculus assume the existence of a field. However, the real numbers are not merely a field. They constitute an ordered field. The field axioms guarantee that the set of objects have the algebraic properties normally associated with real numbers. The ordinal properties are expressed by the following axioms:

7. The real numbers contain a subset, called the positive numbers which also is closed under addition and multiplication.

8. For any real number n, if n is a positive number then the negative of n is not positive.

9. Any real number other than zero is either a positive number or the negative of a positive number.

The real numbers are not merely an ordered field. In addition the set of real numbers is complete. To say that they are complete is to say that they satisfy the following axiom:

10. Every non-empty set (of real numbers) which has an upper bound has a least upper bound.

There are ordered fields which are not complete. For example, the rational numbers are a subset of the real numbers and are not complete. We shall illustrate this fact in explaining the concepts used in expressing axiom 10.

An upper bound of a set of real numbers is a real number which is greater than or equal to every number of the set. For example, if we have the set consisting of the numbers 1, 2, 3, 4, 5 and 6 then the number 7 is an upper bound of this set. It is not however a least upper bound of this set since the number 6½ is also an upper bound of this set but is less than the number 7. The least upper bound of this set is the number 6. Now consider the set of all rational numbers whose square is less than or equal to the number 2. (A rational number is a number which is expressible as a ratio of integers.) As we all know, the square root of two is not a rational number. If the square root of two were a rational number it would be an upper bound of the set in question (since no member of the set could be greater than the square root of two). But any positive rational number not in the set in question would be greater than the square root of two and so the square root of two would be the least upper bound of the set. However, since the square root of two is not a rational number, it is not a rational number which is the least upper bound of the set in question. Nor is any rational number larger than the square root of two a least upper bound of the set since, for any rational number larger than the square root of two there is a smaller rational number which is still larger than the square root of two. Thus, within the set of rational numbers there are sets which have upper bounds but do not have least upper bounds. That is to say, the rational numbers are not complete.

5. The above axioms imply not only that there are real numbers; they also imply that there are infinitely many such numbers. Since numbers apparently are (1) unobservable, (2) neither physical nor mental and (3) universals, many philosophers have been disposed to hold that numbers do not exist. This seems to imply that the axioms stated above are false. Of course, many philosophers who have wanted to say that the real numbers do not exist, have not wanted to say that the above axioms are false and so they have developed theories to explain how this is possible. We shall consider such theories in chapter one. However, before starting this project let us briefly consider the reasons why numbers are neither physical nor mental and why they are universals. (That numbers are unobservable seems hardly contentious. What would it be like to observe zero?)

If numbers were physical then they would have some physical characteristics. For example, numbers would either be moving or at rest, they would have some location in space (or space-time), they would have mass, there might be gravitational forces between numbers. No scientists ever attribute these properties to numbers for the simple reason that they don't really suppose that numbers have such properties. Some philosophers and mathematicians have suggested that numbers might be identified with certain physical objects, e.g., the number one might be a physical line, the number two might be two lines. But there are many serious objections to such a view. For one thing there may not be in existence an infinite number of physical lines (lines of chalk, or string, or steel). For another thing, if numbers were physical lines then imagine the difficulty in distinguishing between a trillion and a trillion and one. Presumably, if numbers were thought of as physical lines then scientists would have devised instruments capable of making such distinctions; but we have no such instruments. No one seriously considers the need to try to devise some such instrument. Subsequently, in our discussion of the theory of mathematical knowledge advanced by the Formalists we shall have further criticisms of this view. Tentatively, at least, we shall accept the proposition that numbers are not physical objects.

That numbers are not mental means that numbers do not exist as a part of or in some human being's mind. To hold that the numbers exist in many minds would imply that there are many zeros, many ones, etc. This is not only surprising, it contradicts the above axioms of real numbers which imply that zero, one, etc., are unique. Of course, one could hold that the numbers exist only in one person's mind. But this view just seems wrong. If we seriously held this view then presumably we would want to know whose mind the numbers existed in; we would want to know what would happen to the numbers when the person whose mind contained them died. But no one is seriously interested in these questions. This strongly suggests that no one seriously thinks that there is only one person whose mind contains the numbers. Subsequently, when we consider the philosophy of mathematics known as intuitionism, we shall consider and criticize this sort of view much more thoroughly. However, at least at present, we shall consider that numbers are not mental.

That numbers are universals seems equally clear. An object is a universal if it can be present in many different objects at the same time. Numbers can be present in different sets of objects at the same time, hence numbers are universals. Let us refer to objects which are universals and which are neither physical nor mental as queer entities. Mathematical statements apparently refer to many sorts of queer entities besides numbers. For example, they refer to sets, functions, polynomials, etc. (Whether these are different from numbers is a question which we shall not consider.) The assertion that there are queer entities, and infinitely many of them at that, is, as we have noted, a statement which many people have wished to deny. We shall now turn to our consideration of the theories people have devised in the effort to account for the truthfulness or at least the usefulness of mathematical statements without accepting the existence of queer entities. We shall present objections to each of these theories which are, in our mind, sufficiently weighty to require their rejection.

6.   Contrary to what we have claimed some philosophers would affirm that we can accept mathematical statements as true without

accepting the existence of queer entities such as numbers. Such philosophers would argue that statements such as

(A) There is a number which when added to 5 yields 0,

do not imply the existence of numbers. According to these philosophers, in arguing that since (A) is true there must be numbers, I am interpreting the expression "There is a number ..." as if it could not be true unless there is an object which is a number and which has the property in question. Contrary to this interpretation of statements such as (A) these philosophers would favor a "substitutional" interpretation of such statements.

Statements such as (A) have the form

(I) $(\exists x)Fx$

where the signs prior to the "Fx" are called an existential quantifier. Statements of this form are sometimes expressed as "There are some Fs" or "There is at least one F." In referring to the "objectual" interpretation of these statements, what is meant is that these statements are to be interpreted as satisfying the following rule regarding their truth conditions:

(Obj) Statements of the form "$(\exists x)Fx$"are true if and only if there is at least one object having the property F.

The substitutional interpretation of these statements rejects the rule (Obj) and replaces it with a rule such as the following:

(Subs) Statements of the form "$(\exists x)Fx$" are true if and only if there is at least one name which satisfies the predicate "Fx."

According to the substitutional interpretation of statements such as (A) we would not be ontologically committed to numbers. Rather we would only be ontologically committed to names of numbers, that is, to numerals. On this interpretation (A) would be true if and only if there is a name which satisfies the predicate "—equals zero when added to 5." There are two questions to ask with respect to the substitutional interpretation. First, is it a satisfactory interpretation of existential statements such as (A)? Does it agree with what we mean when we use statements such as (A)? Are there any cases in which the truth value of statements of the form (A) would be affected by the adoption of the substitutional interpretation? Second, does the adoption of the substitutional interpretation enable us to avoid ontological commitment to queer entities?

It is arguable that the substitutional interpretation of existential quantifications does not agree with what we mean when we use such statements. Consider an alternative example. Suppose my wife asks whether we have any cat food and I say that there is some cat food on the shelf. According to the substitutional interpretation of quantified statements, when I say this it is not implied (by me) that there is any cat food on the shelf. What is implied is that there is a name (presumably the name of a can or box of cat food) which satisfies the predicate "—is cat food and is on the shelf." That is, my answer to my wife would be true if there is a name which satisfies the predicate in question and false otherwise. But I do not think that my answer to my wife would be understood as implying that there is a name which satisfies this predicate. It would be understood as saying that there is a certain object on the shelf. However, this objection is not perhaps all that serious. The defender of the substitutional interpretation can argue that it is not necessary that he provide an account of the meaning of existential statements which preserves their ordinary meaning. What is necessary, he might say, is only to provide an interpretation which preserves the truth-values of factual statements.

But, is it the case that the truth values of factual statements are unaffected by the adoption of the substitutional interpretation? It has been charged by Quine that in a sufficiently well-furnished universe, one in which there is a predicate which is true only of objects which are nameless, that existential quantifications which would be true if interpreted objectually will be false if interpreted substitutionally. He said

> An existential quantification could turn out false when substitutionally construed and true when objectually construed, because of their being objects of the purported kind but only nameless ones.[1]

Dale Gottlieb has attempted to answer this objection by suggesting that there may be no predicate which "is true only of nameless objects."[2] He claims that even if there are nameless objects, there may still be no predicate which is true only of such objects. He suggests that it may be possible to well-order the extension of any predicate by a suitably chosen dyadic predicate and then we could name the objects in the extension of the original predicate by reference to their position in the well-ordering. (A set is well-ordered if every subset under the ordering has a least element. The normal ordering of the positive integers, for example, is a well-ordering.)

This answer to the objection is, at best, rather speculative. Even granting that for any predicate the set of objects in its extension could be well-ordered, it still may be the case that no one knows how to order it. Nobody has been able to produce a well-ordering of the real numbers for example. We can hardly name the real numbers by reference to a well-ordering that has not been specified. Thus, it is not clear that for any predicate with non-empty extension that we can name the objects in its extension.

Gottlieb was trying to argue that there would be no divergence in truth value between existential quantifications interpreted objectually and interpreted substitutionally. He notes that as long as some objects in the extension of a predicate P are named then "$(\exists x)\ Px$" is true no matter how it is interpreted. Similarly if there are no objects in the extension of P then there will be no names which satisfy the predicate P so "$(\exists x)\ Px$" will be false no matter how it is interpreted. The problem of a divergence of truth values between the two interpretations would arise for existential quantifications only in those cases in which all the objects in the extension of a predicate P are nameless. However, there is a problem for universally quantified sentences also. Consider the case of a predicate P which is true of all named objects but false of some nameless objects. In that case "$(\forall x)\ Px$" would be true when interpreted substitutionally but false when interpreted objectually.[3]

So far we have argued that the substitutional interpretation of quantification is not satisfactory since it may have the result that false universal statements are true if interpreted substitutionally whereas true existential statements may be false if interpreted substitutionally. However, we now must turn to the second question which we have raised with the defender of the substitutional interpretation of quantification. The defender of this interpretation, we have suggested, may favor this interpretation in the belief that it will enable him to accept mathematical statements as true without having to accept that there are queer entities. This would be the case if it could be maintained that names which are asserted to exist according to the substitutional interpretation are not queer. However, the defender of this interpretation must hold that names are neither physical objects nor merely ideas in someone's mind for otherwise his view is clearly unsatisfactory. There are names, I would say, of numbers (even of natural numbers) which have never been spoken or written down or even thought in anyone's mind.

Thus the defender of this interpretation of quantification cannot construe names as certain sound waves or mounds of graphite on paper, or traces in some nerve in someone's brain. Further, I think that it is arguable that names are universals rather than particulars since many different objects can have the same name at the same time. Further, names are words; but not all instances of a word need to have been spoken or written. Thus names apparently are queer entities. Apparently the defender of the substitutional interpretation cannot use this interpretation as a way to avoid accepting the existence of queer entities.

NOTES

1. Quine, (74) p. 93.
2. Gottlieb, (24) p. 650.
3. This objection is also noted by Quine, (74) p. 93. For a discussion of the merits of the substitutional interpretation see Putnam, (68) pp. 9–24.

PART I

Chapter One

If-thenism

7.   In this chapter we shall consider a range of views which all agree in regarding mathematical statements as having a conditional form. That is to say, mathematical statements have the form—if such as such is the case then so and so is the case. In virtue of this form, it is alleged that mathematical statements may be true even though mathematical entities such as numbers, functions, etc., do not exist. For example, I referred above to the fact that the real numbers are assumed to satisfy the commutative laws of addition and multiplication. The commutative law of addition may be expressed as the following conditional statement: If a and b are real numbers then $a + b = b + a$. According to the if-thenist, this statement could be true even though there are no real numbers. In this chapter we shall consider two versions of if-thenism corresponding to two different ways of understanding conditional statements. The first version is quite obviously false and is discussed mainly for the sake of completeness. The second version is also, I believe, false but is a more plausible theory. After reading chapter one some people may still find my discussion of the if-thenist position to be incomplete. It may be that the material they expect is in chapter two in which I discuss postulationism.

8.   The first way of supporting the claim that the commutative law (or any other mathematical statement) is true even though there are no numbers is to claim that since the statement is a conditional statement and since conditional statements are true if their antecedents are false, then the fact that there are no numbers is a sufficient condition of the truth of the conditional. Under this interpretation conditional statements are regarded as material implications. The unsatisfactoriness of this position is obvious.

If this if-thenist argument were correct and if all mathematical statements were conditional statements then all mathematical statements would be true. The commutative law would be true but so also would its contrary namely, if a and b are numbers then

$a + b \neq b + a$. Since mathematicians (and most other people) regard the contrary of the commutative law as false with respect to the real numbers, it appears that this if-thenist interpretation of mathematical statements is mistaken. If numbers do not exist, we cannot regard mathematical statements as material implications. However, the if-thenist view which we are considering is committed to the view that numbers do not exist. Since some mathematical statements are false, this version of if-thenism must also be false.

9.   A second version of the if-thenist position is to hold that the if-then relation is not one of material implication but rather of logical implication. That is to say, according to this view, mathematical truths are logical truths. Mathematical statements are conditional statements in which the consequents are logical consequences of the antecedents. This view is combined with the theory of logic which holds that truths of logic are not true in virtue of a correspondence with external reality and so that truths of logic have no ontological implications. If this view of logic were correct and if mathematical truths were all logical truths then clearly mathematical truths would have no ontological implications either. Clearly, this sort of theory is open to two types of criticism. For one thing mathematical truths may not be logical truths and for another thing the theory of logical truths underlying this view may be untenable. We shall briefly discuss the second of these objections first and then move on to the former type of criticism.

The view that logical truths have no ontological implications is sometimes expressed by saying that logical truths are true solely in virtue of their form. This statement is intended to express the idea that statements have a logical form (which need not be identical with their grammatical form) and that in some cases the possession of a certain form is a sufficient condition of the truth of the statement. That is to say, in these cases, the statement is true because of the form it has and regardless of the existence of any objects which may be denoted by the terms of the statement. For example, it may be held that all statements which are instances of the following forms are true solely in virtue of their form: $P \lor \sim P$, $P \supset P$, $\sim (P \& \sim P)$.

But why, we may ask, if we subscribe to the view that logical truths are statements which are true solely in virtue of their logical form, should we accept the claim that logic has no ontological implications. It appears that, on this theory of logical truth, certain statements of logic are true only because there exist certain logical forms and these statements are instances of those forms. Apparently, if we were to accept this theory we would be committed to holding that there are logical forms—entities which are just as queer as numbers or sets. The defender of the view in question is faced with the problem of explaining how it is possible for some statements to be "true solely in virtue of their form" in such a way as to avoid ontological commitment to queer entities. One possible approach might be to say that in affirming that

(A) Some statements are true in virtue of their logical form.

does imply that

(B) There are logical forms.

nonetheless, (A) does not imply that there are queer entities. The defender of the view in question may suggest that logical forms are not queer entities. That is, he may argue for a nominalist theory of logical form or of logical truth. Naturally we are entitled to ask him to spell out his theory for us so that we may judge whether it is adequate.

I suspect that some philosophers who have defended conventionalist views of logic have done so because they thought that a conventionalist theory would enable them to have their logical forms without having queer entities. Such a philosopher might claim that truths of logic are merely instances of sentences which have been affirmed as true by social fiat or convention.[1] For example, it may be claimed that, as a result of such a fiat, the sentence "Everything is either red or it is not red", is to be regarded as true. The philosopher who holds this view may believe that by referring to such concrete entities as sentences and social conventions he will have avoided commitment to queer entities. However, it is not clear that this theory of logical truth will serve the purpose in question. And, furthermore, there are other difficulties.

There are an infinite number (at least potentially infinite) of logical truths. However, it seems most unlikely that there exist the requisite number of social conventions or fiats. Even if we assume that conventions have been adopted at the rate of one per second ever since man appeared on the face of the earth, we would not have as many conventions as there are truths of logic. Rather than assuming that there have been an infinite number of social conventions adopted, it would be more plausible to assume that a finite number of sentences have been adopted by fiat and that the set of logical truths consists of these sentences plus all of their logical consequences. However, if this position is adopted then the defender of this view is no longer giving us an explanation of the nature of logical truths which shows that logic makes no ontological commitments. Instead he would be saying only that the set of logical truths consists of the logical consequences of some set of sentences. This may be true but it is insufficient for its intended purpose. This view, which may be called conventionalism in logic, has been extensively criticized.[2] We shall have more to say about it subsequently.

10. Even if we were to grant that the acceptance of certain principles as principles of logic involved no ontological commitments, serious questions may be raised concerning the claim that mathematical statements, such as the commutative law of addition, are logical implications. That is, we may ask, is $a + b = b + a$ deducible from *a and b are real numbers*? Or alternatively we may ask, is the statement, *if a and b are real numbers then $a + b = b + a$* a logical truth? The answers to these questions are apparently negative. From *a and b are real numbers* we could deduce that $a + b = b + a$ only if further premises were added or if suitable assumptions were made explicit through a definition of the term "real number." The statement *if a and b are real numbers then $a + b = b + a$* is not itself an instance of a valid formula of first or second order quantificational logic. If this statement is a logical truth then it must be possible to transform this statement through the application of definitions of the mathematical terms it contains. Let us investigate possible definitions.

Clearly, not every conceivable definition of "number" would be satisfactory. Turning to commonly used dictionaries to find the "ordinary" meaning of "number" provides a range of possible definitions none of which turn the statement of the commutative law (of of the other axioms of real number theory) into instances of valid formulas. For example, "number" may be defined ostensively as "1,2,3, ...," or it may be defined as "a symbol or group of symbols." More promising perhaps, is to turn to some mathematical theory. Here there are a range of possibilities. Widely known however is the theory in which real numbers are defined as Dedekind cuts. Let us briefly explain this definition.

Within the theory of rational numbers there can be shown to exist certain sets of rationals having the following properties: (i) Such sets contain at least one rational number but not every rational, (ii) If n is a member of such a set and m is less than n then m is a member of the set also and (iii) such a set contains no largest rational. Sets having these three properties are called Dedekind Cuts.[3] As noted above real numbers may be identified with Dedekind Cuts within the theory of rational numbers. So, we may substitute this definition for the term "real number" in the above statement of the commutative law. This yields something along the following lines: *if a and b are sets of rational numbers having properties (i), (ii) and (iii) then $a + b = b + a$.* Clearly, this statement is not an instance of a valid formula of logic. In order to deduce the consequent of this statement from the antecedent additional axioms of rational number theory must be assumed. In particular you must assume that a commutative law of addition is true for rational numbers.

It might be suggested that in turning to the theory of rational numbers for our definition of "real number" we did not turn to a sufficiently fundamental level of analysis. Suppose then that we turned to a more fundamental level, say the theory of sets, to see whether we can find definitions of terms such that the axioms of real number theory can be revealed to be instances of valid formulae of logic. It is readily seen that this result will not be attained even at a more fundamental level. To see this let us consider the axiom which claims that in the real numbers there is an additive identity.

To say that there exists an additive identity is to say that there is a number such that if it is added to any number n the sum of n and that number is n. To translate this claim into set theoretical terms we need suitable definitions of number and of addition. The numbers turn out to be certain sets of objects (which sets depends on which theory of sets is adapted). Addition may be conceived as a function which correlates a number with each pair of numbers. For example, the function might correlate the number 7 with the pair of numbers (2,5). In set theoretical terms functions are themselves identified with sets, in particular, with sets of ordered pairs. The function (set) which correlates 7 with the pair (2,5) would presumably have as one of its elements the ordered pair {7,(2,5)}. The sum is the first element of the pair. Now, to assert that there is an additive identity in the real numbers is to assert that there exists a number which when added to any real number n yields n as the sum. Translated into set theoretical terms, this statement asserts the existence of a set of ordered pairs such that one element of each ordered pair is a pair and that one member of this pair is the set which is identified with the real number 0, and that the first element of every ordered pair in this set is the non-zero element of the pair that is contained in the ordered pair.[4] The ordered pairs $\{5,(0,5)\}$, $\{7,(0,7)\}$, etc., would be members of this function. Clearly, the claim that there exists such a function (set) is not an instance of a valid formula of logic. The function in question has infinitely many members. The statement that it exists is not a theorem of quantificational logic. The statement that such a function exists can be proved in many set theories. But the proof depends essentially on substantive axioms of the set theory. It cannot be proved merely by assuming as axioms the axioms of first or second-order quantificational logic.[5]

11.   In our discussion of if-thenism so far we have assumed that if statements of mathematics are logical implications then they are either instances of valid logical formulas or can be transformed into such instances through replacement of some terms by their definitions. Let us refer to statements which meet either of these conditions as logical truths. So far we have been arguing that the axioms of real number theory are not logical truths. Someone might reject my conclusion because he subscribes to a more inclusive notion of logic than I do. Such a person might hold simply that mathematics is logic and thus that the axioms of set theory (to which I referred as substantive axioms of set theory) are, in reality, principles of logic. He could then conclude that the axioms of real number theory are also logical truths. In reply to this sort of criticism I would reply along the following lines: If one is to speak meaningfully of logical truths then there must be some non-arbitrary distinction between such truths and non-logical truths. If someone includes the axioms of set theory among the logical truths then I want to know how he makes such a distinction. As for me, I can see no more reason for including the axioms of set theory (no matter which one among the current set theories with which I am acquainted) amongst logical truths than I can for including further statements such as those from some axiomatized version of some branch of a modern scientific theory, e.g., biology or physics.

The if-thenist, of course, has a perfect right to ask me to explain or justify my appeal to quantificational logic in my formulation of a criterion for determining whether mathematical principles are logical implications, that is, for my criterion of logical truth. To this question I would reply in the following way: A logical truth, according to my view, should be "true in all possible worlds" in which anything exists. Now, the expression "true in all possible worlds" is metaphorical. However, I believe that the semantical theories developed by Tarski and others provide a reasonably literal interpretation for this expression and so I shall continue to use it.[6] Set theories which are strong enough to provide proofs of the axioms of real number theory are not true in every possible world. For example, there are possible worlds in which there are only a finite number of objects. Or there are possible worlds in which it is not the case that power set of every set of sets exists.[7] Since the axioms of such set theories would be false in such possible worlds they are not, in my view, logical truths.

12.   Let us, at this point, mention one further type of definition of the expression "*a* and *b* are real numbers" (and of related expressions). The expression "*a* and *b* are real numbers" might be understood to mean that *a* and *b* are part of a set of objects which satisfy the axioms of real number theory. That is to say that *a* and *b* are part of a set of objects which are closed under addition and multiplication, etc. Then indeed, the statement "if *a* and *b* are real numbers then $a + b = b + a$" would be an elementary logical implication. Have we at last found a satisfactory way of showing that the truths of mathematics have no ontological implications? In the next chapter we shall explain this view of mathematics in more detail and then consider further criticisms.

13.   Conclusion: In this chapter we have considered the view that mathematical statements have no ontological implications because they are all statements of logical implications. In criticizing this view we have considered statements of real number theory. They are not statements of logical implications in themselves and further they are not transformed into such statements when mathematical terms are replaced by their definitions.

NOTES

1. Views of this sort have been expressed by Hans Hahn in (25) and by A.J. Ayer in (1) where he says of logical and mathematical propositions that "They simply record our determination to use words in a certain fashion.", p. 84, and by Carnap in (10).

2. For extensive criticisms of this view see Quine, (71) and Hilary Putnam, (66) and (65).

3. Walter Rudin, (80) p. 3.

4. A minor modification of this last statement is necessary to include the sum $0 + 0$. Clearly, our argument is not affected by this change.

5. First-order quantificational logic may be explained as follows: The formulae of first-order quantificational logic contain two types of symbols. One type are commonly referred to as individual variables and are interpreted by reference to individuals. The other type are normally called predicate letters. They are interpreted as specific properties or relations. In first-order quantificational logic the range or scope of the quantifiers is restricted to the individuals in the universe of discourse with respect to which the formulae are interpreted. In second-order logic there are also quantifiers which range over the properties or sets of the universe of discourse.

By "substantive axioms of set theory" I mean such axioms as the axiom of infinity, or the comprehension axioms which are not necessary for "proofs" of any of the valid formulas of quantificational theory, that is, for "proofs" of any formulas which are true in all non-empty domains. There are, of course, a number of distinct set theories such as the theory of Principia Mathematica or the Zermelo–Frankel set theory. For an introductory account of these see (26). In addition one may consult (18) or (73).

6. Roughly, a possible world is a non-empty universe of discourse (or domain). There may exist in such a universe individuals or sets or properties or relations.

7. The power set of a set is the set of all subsets of the given set. For example the power set of the set whose elements are the numbers 1,2, and 3 consists of the set whose members are the singletons $\{1\}$, $\{2\}$, $\{3\}$, plus the doubletons, $\{1,2\}$, $\{2,3\}$, $\{1,3\}$, plus the set $\{1,2,3\}$ and finally the null set.

Chapter Two

Postulationism

14.   The last suggestion regarding the definition of "real number" brings us quite close to a theory expressed by Bertrand Russell in his book *Principles of Mathematics*. Russell claimed in that book that pure mathematics consists solely of propositions of the form "*p* implies *q*."[1] This view is not the same as that we have been considering in chapter one. In that chapter we were considering the view that mathematical principles such as the axioms of real number theory are themselves statements of valid logical implications. We have found that view to be untenable. The view we are here considering, which we shall call postulationism, accepts the fact that statements such as the above axioms are not logical truths.[2] It holds rather that pure mathematics consists in deducing by purely logical inferences the consequences of freely chosen postulates. For example, Russell claimed that with respect to Euclid that "What pure mathematics asserts is merely that the Euclidean propositions follow from the Euclidean axioms."[3]

Postulationism may also be expressed as the claim that the axioms of a mathematical theory constitutes a set of specifications which define a structure that the mathematician wishes to study. Thus, the axioms of real number theory may be said to "define" the real numbers. The real numbers are any set of objects which happen to satisfy these axioms. Whether any such objects exist, it is alleged, is of no concern to the mathematician.

The postulationist, by calling the axioms definitions, wishes to draw attention to a contrast between statements of definitions and other declarative statements which are accepted as true. Often, as noted in the introduction, the acceptance of declarative statements as true carries the implication that there are objects and that the truth of the statements is due to their corresponding to the objects in question. By calling the axioms definitions, the postulationist wants to say that even though the axioms are studied, there is to be no implication that there are objects corresponding to these statements. This view was expressed by Poincaré as well as by Russell. Poincaré claimed (with respect to geometry) that the axioms are "definitions in disguise."[4] Poincaré did not think that this was true of all mathematical disciplines. He specifically excluded arithmetic.

15.   Can we accept postulationism and so avoid ontological commitment while accepting mathematics as a body of knowledge? We should note immediately of course that even if postulationism is correct with regard to "pure" mathematics, it will not enable us to avoid ontological commitments as soon as mathematical principles are applied in the solution of problems of other sciences or everyday life. This can be made clear by considering a simple example involving only the application of the algebraic properties of numbers. To see this let us consider briefly a simple problem:

> There are two solutions of iodine, one of strength 8% and the other of strength 24%. How much of each solution should be taken to obtain 16 oz. of a 12% solution of iodine?

To solve such a problem the student says: Let x represent the amount of the 8% solution (in ounces) and let y represent the amount of the 24% solution. You are told that

(a) $x + y = 16$

and also that

(b) $.08x + .24y = .12(16)$

From these two equations you can easily solve for x and y. For example, using the fact that the real numbers are closed under addition, you can add $(-y)$ on each side of equation (a) thus obtaining

(c) $x = 16 - y$

from which you can infer by the rule of substitutivity of identity

(d) $.08(16 - y) + .24y = .12(16)$

Then using the fact that the reals are closed under multiplication you can infer that

(e) $.08(16 - y) + .24y = 1.92$

etc.

How is this process of reasoning to be analyzed? It seems reasonable to assume that steps (c), (d) and (e) are to be regarded as consequences of steps (a) and (b). But, if these consequences are to be considered as following validly by principles of (first-order) logic, then clearly the principles of closure under addition and multiplication of real numbers as well as the principle of substitutivity of identity are also assumed. Thus, this process of reasoning may be regarded as consisting of the following statements: We know that there are true statements of the form (a) and (b) and we want to find out what these statements are. Thus, we must find out the values of x and y. Using (a) and (b) and certain other truths (e.g., closure of the real numbers under addition and multiplication) we can infer that the values of x and y must be such and such and so and so (the problem of finding x and y we leave to the reader). In any case, the mathematical principles assumed here are not hypothetical, e.g., it is assumed categorically that the real numbers are closed under addition and multiplication. That is, we would not know that $x = n$ and $y = m$ (whatever n and m turn out to be) unless we also know not merely that if x and y are real numbers they are closed under addition. We must know that when we add $(-y)$ to obtain line (c) above that the sum of the numbers $(x + y)$ and $(-y)$ exists. Presumably we know this fact, even though we don't know the values of x and y, because we know that x and y are real numbers and that the sum of any two real numbers exists. Given the pattern of reasoning employed here, we may conclude that since we accept the statements of the solution to the problem as true, then we also accept as true the statements from which the solution was deduced. Even if postulationism is true as a theory concerning "pure" mathematics, it provides no reason for concluding that when we apply mathematics we do not imply that queer entities exist.

16. But we must also ask whether postulationism is true as a theory concerning the nature of pure mathematics. According to postulationism, the pure mathematician specifies some structure which he wishes to study by adopting a certain set of axioms. The structure is whatever the axioms say that it is. The axioms define the structure. Study of the structure consists simply in deducing whatever consequences can be deduced (by pure logic) from the axioms.

It appears then that postulationism is to be considered as a descriptive theory. What it purports to describe is the activity of the "pure" mathematician. But the description that it gives is obviously unsatisfactory. For one thing the activity of deducing whatever consequences can be obtained from a given set of axioms is far too trivial for anyone to spend much time doing it. Infinite numbers of consequences which could be drawn from a set of axioms are not. For example, if we represent a set of axioms by the letter A and an infinite number of other statements by the symbols $Q_1, Q_2, \ldots$ then we immediately could deduce from A all of the following theorems: if Q_1 then A, if Q_2 then A, ... But if one turns to works on mathematics one notes that there are enormous numbers of theorems of this sort which could have been deduced but which the mathematicians have not deduced. The least we can say is that postulationism owes us an explanation of why mathematicians apparently fail to make so many obvious deductions. Why, for example, don't books on the theory of real numbers start out by first listing the axioms for real numbers and then deducing such statements as "If Dodo's are blue then the real numbers are closed under addition." "Either elephants have wings or there exists an addititive identity", etc.

But postulationism doesn't seem to offer a correct (even though incomplete) description. The work of mathematicians often does not consist of deducing consequences from explicitly stated axioms. This was the case in Euclid's time and it is still so. For example, in a recent monograph on convex figures, the author commences by presenting several definitions, e.g., of convex figure and bounded convex figure. He then proceeds directly to the proof of a theorem. No axioms are stated though several axioms are clearly taken for granted.[5] But, if the postulationist were correct, the axioms should be explicitly stated. If they are not stated then how can we check to see whether the deductions are valid?

If the last cited work is looked at, it does not even appear as if the author is merely deducing consequences from axioms. The author starts by giving the following definition: "A figure is said to be convex if it entirely contains all segments connecting any two of its points." ... "A convex figure is said to be bounded if it can be imbedded in a circle of finite radius; otherwise, unbounded." He then gives some examples of convex figures and discusses one-dimensional convex figures, e.g., a straight line, and two dimensional convex figures, e.g., triangles. No axioms are stated, albeit the author does then proceed to state and prove a theorem, namely, "If a convex figure Q contains three points A, B, C, which do not lie on one straight line, then Q contains all of the triangle ABC." This theorem is supported by the following argument: "In fact, since Q is a convex figure containing the points, A, B and $C \ldots$ it must contain all three sides AB, AC and BC of the triangle ABC. The triangle *ABC* can be covered by segments *AE* joining the vertex *A* to points of the opposite side *BC*. Since the ends of the segments *AE* belong to the figure *Q*, each of the segments *AE* belongs entirely to *Q*. Hence all of the triangle *ABC*, being covered by such segments, belongs to *Q*."[6] It appears that the author wishes us to picture to ourselves a certain figure and to "see" that the statements which he makes concerning the figure are correct. Presumably in his own thinking about complex figures this is what he has done. To say that he has adopted axioms and deduced theorems from the axioms hardly seems to be a correct description of what the mathematician has been doing. It would be far more correct to say that he has been studying geometric figures of a certain sort, namely plane convex figures, and trying to discern some of their characteristics. For example, the reader supposedly can see that the side of the triangle ABC can be covered by the line segments AE as described. Let us go into this matter at greater length.

The mathematician, from his earliest mathematical experience, becomes familiar with various types of objects, such as the natural numbers, or with various structures that are manifest in sets of such objects, such as the structure of a field. For example, as one learns mathematics in school one learns eventually that all equations of certain forms have solutions. One learns that certain operations, e.g., addition, can always be performed. One learns that every real number has a square root. Through his experience in this way the mathematician acquires concepts of various sorts of mathematical objects. He may eventually come to realize that his knowledge of these objects is unsystematic. He may wonder whether the arguments he has accepted as proving that certain statements are true are satisfactory. He may suspect that at some point in his arguments he has included a false statement or begged a question. In the effort to discover whether his arguments are satisfactory he may begin to try to formulate it in a systematic way. For example, he may distinguish between the statements for which he has proofs and the statements which he takes as true but which are unproved. To say that the mathematician has simply adopted some axioms and deduced consequences from them, ignores the fact that what he has been doing is trying to bring system into a set of statements which he regards as truly describing objects of certain sorts.

If we take the view that mathematics is simply the activity of deducing consequences from axioms then certain questions which are of great interest to some mathematicians become unintelligible. If the mathematician merely adopts axioms and deduces consequences then what sense can we make of the mathematician's concern that his axioms be complete and categorical? Let us briefly explain these notions of completeness and categoricalness. They help us to bring out inadequacies in the postulationist theory.

To explain the notion of categoricalness let us consider an example. Suppose that the mathematician is trying to develop in an orderly manner his knowledge of the so-called counting numbers (or natural numbers), namely of 0,1,2,3, ... He may attempt to formulate his knowledge by deducing theorems from the following set of axioms:

(1) 0 is a number
(2) 0 is not the successor of any number
(3) Every number has a successor
(4) No two numbers have the same successor
(5) If a property is true of 0 and also true of the successor of a number when it is true of the number, then the property is true of all the natural numbers.

These axioms are called Peano Postulates. As has long been known, the Peano postulates do not characterize the natural numbers so precisely that no other set of objects satisfies all of the axioms. Another set of objects which satisfies the Peano Postulates is the set of even numbers. The Peano Postulates are not categorical.

We may draw an analogy between the problem of the mathematician who attempts to discover a categorical set of axioms and a biologist who is trying to define precisely a group of organisms which he has been studying. Suppose that a biologist thought that he had come up with a satisfactory specification of all and only the essential features of a group of fish only to discover that these same features were manifest by some other group of organisms such as worms. He would reject his definition as unsatisfactory. Just as the biologist seeks for a more satisfactory definition of the group he has in mind, the mathematician desires his postulates to be categorical. (For further discussion of the concept of a categorical axiom set see Foundations of Set Theory.[7])

Failure to characterize uniquely a given set of objects is only one sort of inadequacy of an axiom system. A set of axioms is even more clearly inadequate if there are truths concerning the objects in question which cannot be derived from the axioms or if there are statements which are known to be false but the falsehood of which cannot be deduced from the axioms. Let us say that a set of axioms is complete if all statements concerning a set of objects, which are true, can be deduced from the axioms. But, what sense can be made out of these concerns (the concern to have one's axioms complete and categorical) from the postulationist point of view? If the only mathematical truths are statements of logical implications, i.e., statements of the form "Axioms logically imply Theorem", then there are no truths which a given set of axioms fails to imply and so no axioms can be incomplete in the above sense. If the mathematician is merely *defining* structures by adopting axiom sets rather than trying to give a description which fits a set of objects uniquely, then to claim that some axioms are either categorical or non-categorical would be meaningless and we would hardly expect to find statements of the following sorts in distinguished mathematics texts:

> "A categoric postulate set is a sort of arch of triumph. When we are able to write such a postulate set for a particular mathematical structure, this means that we have a complete understanding of its essential properties."[8]

17.   The view that we have been calling "postulationism" was criticized by Quine for a different reason.[9] Consider, for example, the following statement which many people regard as a mathematical truth:

(1) For any real numbers a and b, $|a + b| \leq |a| + |b|$

According to postulationism, this statement is not a mathematical truth. But there is a corresponding mathematical truth namely

(1') The axioms of real number theory imply (1)

According to Quine, while postulationism makes what sounds like an interesting claim, namely that truths of mathematics are in fact truths of logic, all that postulationism really amounts to is a redefining of the scope of the term "mathematical truth" in such a way as to exclude truths most people consider as mathematical truths. Postulationism contains no theory concerning the way in which we come to know truths such as (1) and so, in fact, is not as interesting a theory as it may appear to be. Nor does postulationism provide us with any grounds for thinking that mathematical statements such as (1) can be accepted as true without any ontological commitment.

Quine noted that in a similar manner (to that adopted in postulationism) one could support the view that truths of sociology are truths of logic. All one would need is a set of statements to serve as axioms. Assuming that the truths of sociology in question were entailed by these axioms one could then claim that all truths of sociology were in fact statements of logical implications analogous to (1'). Nor will it do to reply to Quine that there are no acceptable sociological axioms. For, even if there were such axioms in sociology, we would not consider that the postulationist approach showed that truths of sociology were merely truths of logic. In a subsequent chapter, when we shall be concerned with the question of how mathematical statements such as (1) can be known, we shall have further criticisms to make of this view.

18.   Conclusion: In this chapter we have considered the view that mathematics consists solely in the deducing of consequences from hypotheses or alternatively in the study of the logical consequences of structures defined by sets of postulates. We have seen that this view does not enable us to avoid ontological commitment to mathematical entities when we apply mathematical principles in science. Further, this view does not seem to yield a correct description of what mathematicians do nor of the class of mathematical truths.

NOTES

1. Bertrand Russell, (84) p. 3 and the essay "Mathematics and the Metaphysicians" in (82).
2. The view we are calling *postulationism* was called *if-thenism* by Putnam in (67).
3. Russell, (84) p. 5.
4. Poincaré, (60) p. 50.
5. Lyusternik, (49).
6. *Ibid.*, pp. 1–3.
7. Fraenkel, (18) pp. 297 ff. The definition of 'categorical' used here is somewhat simpler than the definition commonly used. According to the more common usage a postulate set is said to be categorical if all of its models are isomorphic. To utilize this definition of 'categorical', we must consider the postulates in question as formulae rather than as meaningful statements. It has been proved that, so construed, the Peano axioms are non-categorical, though the example given in the text does not show that this is so, i.e., does not show that there are models of the Peano axioms which are not isomorphic to each other.
8. Moise, (56) p. 386.
9. Quine, (71) p. 327.

Chapter Three

Mathematical Principles as Analytic

19.   Many philosophers have maintained or presupposed a view according to which all true statements (or all factual statements) may be classified as either analytic or synthetic. One advocate of this view, Carl Hempel, explains the distinction between analytic and synthetic truths as follows: According to Hempel, analytic truths are

> true simply by virtue of definitions or of similar stipulations which determine the meaning of the key terms involved.[1]

Truths which are not analytic are called synthetic. Let us explain Hempel's view a bit more carefully with reference to an example. Consider the arithmetical truth that $3 + 2 = 5$. The terms contained in the statement of this truth are '3', '2', '+', '=' and '5'. According to the view in question these terms have meanings attached to them. That is, there is some meaning attached to the sign '2', etc. And further, according to the view we are considering, we can tell that the statement that $3 + 2 = 5$ is true simply by considering the meanings which are attached to the signs that it contains.

A second claim made by advocates of the view we are here considering is that analytic propositions have no factual import. According to A.J. Ayer, analytic propositions do not

> provide any information about any matter of fact. In other words, they are entirely devoid of factual content.[2]

In claiming that mathematical propositions are devoid of factual content, philosophers such as Ayer and Hempel, were making two claims. First, they were saying that mathematical propositions are not confirmable or verifiable by reference to sensory observations. In other words we do not know propositions such as "$3 + 2 = 5$" because of anything that we have observed. Such statements are neither confirmable nor refutable by reference to observations. There are no sensory observations which anyone needs to make in order for men to know that $3 + 2 = 5$. Second, they were saying that mathematical propositions have no ontological import. That is, even though statements such as "$3 + 2 = 5$" are true, this does not mean that anything exists. Such statements would be true even if nothing existed. In other words such statements do not correspond to any external reality. Nor are they true because of such correspondence.

Actually, in the work *Language, Truth and Logic* Ayer did not present the definitions for us to inspect in order to determine the truth of some mathematical propositions. Hempel in the work cited above, does present some definitions which he alleges are sufficient for determining the truth of arithmetical statements. However, the fullest expression of this theory is to be found in the work of Rudolph Carnap.[3] Subsequently we shall present and critically discuss Carnap's work on this matter at some length. Prior to undertaking such discussion, however, it will be convenient to discuss some of the issues which this view has raised.

20.   The view that arithmetical and other mathematical truths are true because of the definitions of the terms in the statements expressing them may appear plausible if one considers simple truths such as "3 + 2 = 5." It may be argued that the sign '3' is only an abbreviation for some such mark as "111" and that '2' is an abbreviation for the mark "11." Furthermore, it may be suggested that as a result of the definition of '+' we are to understand the sign "3 + 2" as signifying the mark obtained by adjoining the mark signified by '2' to the mark signified by '3'. Similarly the sign '5' is understood, it is argued, to signify the mark "11111" so that the expression "3 + 2 = 5", it is alleged, is merely a convenient way of writing "11111 = 11111."

At this point, I believe, we are simply supposed to agree that "3 + 2 = 5" is "true by definition." But, what if we are not yet prepared to agree, and instead point out that all that the above definitions have accomplished is the transformation of one statement into another statement? Have we any reason for holding that "11111 = 11111" is "true by definition?" After all, we have not yet been given the meaning of the mark "11111." To this question we may expect the response that it does not matter what "11111" signifies. We may be told that in virtue of the meaning of the sign '=' we know that as long as the same term appears on both sides of the identity sign the resulting statement is true. Thus we must ask, what is the meaning attached to the identity sign and if we agree that statements such as "11111 = 11111" are true simply in virtue of the meaning of identity, does it follow that such statements are devoid of ontological import?

In logical theory the identity sign is commonly explained by reference to the following two laws of identity, namely

(i) $(\forall x)\ (x = x)$ and (ii) $(\forall x)\ (\forall y)\ ([x = y\ \&\ Fx] \supset Fy)$

The first of these laws says that everything is identical with itself. The second says that for any objects x and y, if x and y are the same thing and x has the property F then y has the property F also. The second of these laws can be understood as a principle of second-order quantificational logic since, in effect, we understand it as if there were a universal quantifier in front of the whole expression which binds the predicate letter F. In order words, we understand it as if the letter F were a variable ranging over properties.

But, do these laws of identity hold in virtue of the meaning of identity? Those who would hold that these laws hold in virtue of the meaning of identity present us with the following alternatives. Logical principles such as the laws of identity are either (1) laws of thought (2) empirical generalizations or (3) true in virtue of the meaning of terms. They then argue that such principles are neither laws of thought (since people don't necessarily think logically—they sometimes reason invalidly) nor empirical principles (since logical principles are allegedly not refutable by counter-examples) and so they must be true in virtue of the meaning of the terms of the statements by which they are expressed.

This conclusion may be questioned. If, as Quine has suggested in a number of places, logical principles are very basic principles of science, they will be relatively immune to refutation by counter-example. But they will not be absolutely immune.[4] Conceptual developments in other scientific principles and observational data may require the modification of logical principles. (Putnam has gone further in this direction and argued that developments in quantum theory indeed require the modification of our principles of logic.[5])

If we follow Quine and Putnam we will conclude then that we are not forced to accept the conclusion that logical principles are statements true in virtue of the meaning of their terms and consequently that the acceptance of logical principles is not necessarily devoid of ontological implications. However, even if we agree that the laws of identity (and other logical principles) are true in virtue of the meaning of their terms, we need not be forced to the conclusion that such principles have no ontological import. The explanation of the laws of identity given above was done with reference to the higher order logic of quantification theory. But how are these quantificational statements to be understood? In the second principle we noted that we have, in effect, a quantifier which ranges over all properties. Thus, this statement is true only if there is a non-empty domain of properties.[6] This suggests that while the logical principle may be true in virtue of the meaning of identity (or of other logical expressions), the statements explaining the meaning of identity have ontological implications.

Advocates of the philosophical view we have been criticizing, e.g., Ayer, Hempel, Carnap and others, subscribe to the claim that if a statement is true by virtue of the definitions of its terms, then the statement has no factual or ontological import. But, as we have seen there is reason to question this claim. What we have just argued is, in effect, that the definitions of the terms of the logical theory presuppose another theory which in turn has ontological implications.

21.   We argued above that we are not compelled to say that arithmetic truths are true by definition of the terms of the statements in which they are expressed. Further support for this view is obtained when one considers the ways in which these statements are learned. One does not come to know that $2+3=5$ by looking up the definitions of these terms in some sort of a dictionary. Clearly, the way that everyone first learns these statements involves counting appropriate sets of objects. For example, one counts a set having two objects and then a disjoint set having three objects and then counts the union of these two sets.

Those who accepted views such as those of Hempel and Carnap would respond to the above argument by claiming that talk of counting in the above context is irrelevant. They would say that such talk confuses psychological or genetic matters, such as matters relevant to how one learns a truth, with epistemological matters. For example, Ayer has said

It is obvious that mathematics and logic have to be learned in the same way as chemistry and history have to be learned ... What we are discussing however, when we say that logical and mathematical truths are known independently of experience, is not a historical question concerning the way in which these truths were originally discovered, nor a psychological question concerning the way in which each of us comes to learn them, but an epistemological question.[7]

At this point it is natural to ask how we are to distinguish epistemological matters from these other psychological or historical matters? And to this, Ayer suggests the reply that, the epistemological question concerns the way in which such propositions are "validated." Perhaps this suggestion can be understood in the following way. To see how a proposition is validated is to see what reasons scientists offer in evidence for the proposition. With regard to many propositions, it is clear that the ways in which we come to learn that they are true may not involve learning the reasoning through which they are "validated." For example, we may learn that twelve times twelve is one hundred and forty-four because someone tells us that this is true and not because we have studied the arithmetical proof. In other cases we may learn that some principle is true through a combination of factors which may involve (1) witnessing examples involving the application of the principle and (2) being told or reading that the principle is true, etc., where the evidence considered as necessary to prove the proposition is considered too extensive or too complex for us to grasp. Thus, we can grant Ayer and Hempel the claim that there is a legitimate distinction between the experiences through which an individual learns a particular proposition and the evidence which is considered as proving or validating that proposition. But, that does not show that considerations relating to the ways in which arithmetical propositions are learned are totally irrelevant to the question of how they are proved.

If Ayer and Hempel are correct in their view, then we should have to say that until the work of Peano and Boole and others was completed in the nineteenth century, while many people had learned arithmetic, nobody knew that arithmetical principles were true since nobody had learned the appropriate definitions. In other words, on this view we should say that mathematicians such as Leibniz, Newton, Descartes, Archimedes, etc., did not know such truths as that $2 + 3 = 5$ since Peano (or even more Frege and Russell) hadn't yet given *the* definitions of the requisite terms. Surely, this is an absurd consequence of the theory we are considering. Peano and others surely knew that some arithmetical truths were true when they were very little even though nobody knew the definitions which were discovered later.

Let us distinguish two ways in which a proposition can be learned. In the first way, a person learns that a proposition is true, when he learns the evidence which confirms or proves the proposition and learns that this proposition is confirmed or proved by the evidence in question. The evidence in question is the evidence available to scientists. And, let us note that, if no one is aware of this evidence then no one knows that the proposition is true (though people might, of course, believe that is true even though they do not know it). Learning in the second way does not involve consideration of the evidence available to scientists but occurs as a result of being told or reading that a proposition is true. Clearly, we may agree that considerations relating to learning in this second way may be entirely irrelevant to describing the reasoning or evidence through which a proposition is validated. And, when Ayers refers to a distinction between epistemological and psychological matters, we can see that he is correct if he is referring to learning in the second way. But his claim is not correct if it is understood to refer to learning in the first way. For, in regard to learning in the first way, the way that one learns a proposition is also the way that it is validated.

To maintain that consideration of the way in which propositions of a certain sort are learned is totally irrelevant to the way in which propositions of that sort are validated is, in effect, to maintain that all learning of propositions of that sort is learning of the second kind. Thus, if Ayer maintains that considerations of the way arithmetic is learned is irrelevant to the way that arithmetic is validated then he must be maintaining that all learning of arithmetic is of the second kind. But surely this is mistaken. Not all cases of learning of a proposition can be learning of the second sort. You cannot learn that a proposition is true through reading or hearing it unless someone has learned that it is true through consideration of the evidence which supports it. Learning that a proposition is true in the second way may be called parasitical upon learning that a proposition is true in the first way.

Confusion on this matter may be a result of failing to heed another distinction. We should distinguish between learning that a proposition is true and merely learning to say it or read it or even to think it to oneself. Clearly, one can learn to say or read a proposition and yet not know that it is true. When Ayer claimed that the way a proposition is learned is logically independent of the way in which it is validated, he may have been thinking of learning a proposition not in the sense of learning that the proposition is true, but in the sense of learning to say it or read it. Now he may be correct in that the way one learns to say or read some mathematical proposition is logically independent of the way that the proposition is validated. But if we were to accept Ayer's view that the truths of mathematics are validated through reference to definitions provided (or discovered) by Peano, Russell and others, then we would have to say that prior to the completion of the work of these men while many people had learned to say or read arithmetical propositions, nobody knew that they were true. Surely, this is an implausible view. Many people have known many arithmetical truths and not merely how to say or read such truths. This knowledge existed long before the work of Russell and others was completed. Since this knowledge could not all have been obtained in the second way noted above, at least some people must have learned arithmetical truths in the first way. It is not implausible that the picture of learning arithmetic through counting sets as described above constitutes part of this kind of learning.

To summarize this section let us say that according to Ayer, no one could have learned in the first way that arithmetical propositions are true through counting sets of objects (since, on his view, that is not the way such propositions are validated). We have tried to refute this view, by deducing from it an absurd consequence, namely, that since the definitions by which arithmetical propositions were allegedly validated were not known until recently, it follows on Ayer's view that no one knew that any mathematical propositions were true until recently and that even the mathematicians and logicians who developed the definitions in question did not know that the arithmetical propositions were true prior to their writing advanced works in logic. They did not know these arithmetical propositions were true because, on Ayer's view no one could have learned them (through the first kind of learning). If no one could have learned them in the first way then no one would have known them.

22.   In defending their view that arithmetical propositions are analytic, Hempel and Ayer have argued that such propositions are logically immune from refutation through observed counter-instances. Hempel gave the following example:

> We place some microbes on a slide, putting down first three of them and then another two. Afterwards we count all the microbes to test whether in this instance 3 and 2 actually added up to 5. Suppose now that we counted 6 microbes altogether. Would we consider this as an empirical disconfirmation of the given proposition, or at least as a proof that it does not apply to microbes? Clearly not; rather, we would assume we had made a mistake in counting or that one of the microbes had split in

two between the first and second count. But under no circumstances could the phenomenon just described invalidate the arithmetical proposition in question. ...[8]

Ayer makes a similar claim. He says that "one would adopt as an explanation whatever empirical hypothesis fitted in best with the accredited facts."[9] He claims that in no circumstances would one adopt the explanation that $3 + 2$ is not equal to 5. A similar view of arithmetic was expressed by C.I. Lewis, namely that we are prepared to maintain the truth of arithmetical and indeed of all mathematical truths no matter what happens in nature. Lewis is considering an argument offered by J.S. Mill in accordance with which we are to suppose a powerful demon who could arrange that when we added two things to two things the result would turn out five. According to Lewis, even if we found on counting two things and two things and then counting the sum we found five things we would not reject the arithmetical law that $2 + 2 = 4$. Lewis maintains that the arithmetical law is analytic, that is, a consequence of our definitions. If Mill's demon existed we might have to revise our physics or chemistry or we might have to suppose that we were undergoing some strange opitcal illusions, but we would not change our arithmetic.[10]

But, how strong is this argument? Is it really the case that under no circumstances would we be willing to give up our arithmetical principles? Suppose we performed the experiment described by Hempel and got the result he describes. And suppose that this was repeated many times. Suppose further that we began to consider some of the alternative hypotheses that have been suggested. Since the microbes might subdivide at any point we have a film of them running continuously which we scan and see that in fact no subdivision occurs. Suppose that we have several different observers and suppose that all known circumstances under which optical illusions occur are eliminated. Suppose finally that the two sets of microbes are not physically combined but that, even with all these precautions, when we count the microbes the second time we find that the sum is six. In these circumstances would we agree with Hempel that we would still maintain that the arithmetical proposition $3 + 2 = 5$ applies to microbes?

Of course, we are not logically compelled to modify our arithmetic in the circumstances that I have described. We can continue to maintain that the arithmetic is true. We can continue to search for alternative biological, physical, optical, etc., hypotheses to explain the phenomenon in question. But, suppose that each time a plausible biological, etc., hypothesis is suggested as an explanation for the deviant phenomenon, the hypothesis is found to be unsatisfactory by other considerations, e.g., plausible chemical hypotheses which explain the phenomenon and save the arithmetic always are found to conflict with other well-confirmed hypotheses. Couldn't we conclude that the most plausible explanation for the phenomenon in question involved rejecting the applicability of traditional arithmetic to the biological phenomenon in question? In fact, let us suppose that in addition to the failure of all suggested hypotheses of the above sort a plausible explanation of the phenomenon is proposed which (a) does not conflict with well-confirmed theories in physics, chemistry, etc., in any known way and (b) involves the rejection of traditional arithmetic and the use of a "queer arithmetic."[11] If this supposition were realized it would seem that we not only could conclude that the traditional arithmetic is false for microbes but that we should draw some such conclusion.[12] At any rate, the suggestion that arithmetic is immune from counter-examples does not appear to be well-founded. It might indeed be the case that we would be rationally warranted in rejecting ordinary arithmetic.

23.   While holding that mathematical principles cannot be refuted by empirical counter-examples, some philosophers have allowed that we might be led to modify or reject mathematical principles on "grounds of conformity to human bent and intellectual convenience."[13] This suggests, for example, that while we cannot be logically compelled to give up the claim that $3 + 2 = 5$ or even to allow that this claim is not true of microbes, we may find it convenient to give up this claim. Pragmatic considerations may lead us to favor one set of logical or mathematical *rules* in favor of another. I have italicized the term "rules" to call attention to the fact that on the view in question mathematical principles are rules rather than descriptive statements. Philosophers who held this view apparently thought that if mathematical principles are rules rather than descriptive statements then mathematical principles would have no ontological import. In criticizing this view I shall argue that while mathematical principles may indeed be construed as rules, this does not show that such principles have no ontological import. Let us consider this view in more detail.

Advocates of the view that mathematical principles are analytic fell into two camps with respect to a distinction concerning the significance of analytic propositions. In one camp are philosophers such as A.J. Ayer who hold that analytic propositions are statements whose truth is determined by rules of common usage of terms. Ayer says that analytic propositions "call attention to linguistic usages."[14] This isn't to say that mathematical propositions actually describe the way words are used. Ayer doesn't say that. What he says is that the patterns of usage determine which propositions are analytic. In the other camp are philosophers such as C.I. Lewis and Ernest Nagel who held that mathematical propositions are rules which we adopt and which prescribe the way that we are to use language or the way that we are to think. Lewis said that laws of logic are "principles of procedure, the parliamentary rules of intelligent thought and speech."[15] Nagel, in discussing the Aristotelian principle that "the same attribute cannot at the same time belong and not belong to the same subject in the same respect" said that "*the principle is employed as a criterion* for deciding whether the specification of the attribute is suitable ..."[16] He claims that the principle functions "as a norm or regulative principle for introducing distinctions and for instituting appropriate linguistic usage."[17] Such rules are adopted, according to Lewis and Nagel, in order to facilitate intelligent communication. According to Lewis, the principles of arithmetic are true in any possible world. This is because such principles are analytic consequences of rules having to do with counting.[18] The implication of this remark is that arithmetic principles and the "analytic" judgments they determine have no ontological or factual meaning. Nagel also drew this conclusion. He claimed that the ontological interpretation of logical and mathematical principles is unwarranted. He said "The interpretation of logical principles as ontological invariants seems ... to be an extraneous ornamentation. ..."[19] and also that the role of "logico-mathematical" principles can be understood without "the invention of a hypostatic subject matter for them."[20]

Underlying this view there is a distinction between sentences which express rules (or norms or ideals), that is, which prescribe some mode of behaviour, and sentences which state facts about the world, that is which assert that something exists or that some existent thing has certain properties or stands in certain relations to other things. Sentences of the latter sort are called descriptive sentences. The former are called prescriptive. A sentence such as "The cow is brown" is descriptive, whereas the sentence "Close the door" is prescriptive. Mathematical sentences, according to the view in question, superficially appear descriptive but are, in reality, prescriptive. That they are really prescriptive may be discerned by a study of the role they play in discourse. Consider, for example, the Aristotelian principle of contradiction mentioned above. According to the prescriptivist, this principle is not really about attributes or entities. It is not a supremely general descriptive principle—a principle about all possible beings. Rather this principle instructs us that we are not to apply an attributive term of an object in one respect and also deny the term of the same object in the same respect. Alternatively, if we both apply and deny the same term to an object then, the principle tells us that the term cannot have been applied in the "same respect."

That logical principles may be interpreted as prescriptions rather than as descriptions is not implausible. That mathematical principles should be interpreted as prescriptions rather than as descriptions is less plausible in my view. Consider, for example, the axiom of choice.[21] One way of stating this principle is as follows: In any set S of disjoint sets not containing the null set, there exists a function f whose domain is S and such that for each member m of S, $f(m)$ is a member of m. The function f is called the choice function. The axiom is explicitly existential in character. It claims that in any set of sets satisfying a certain condition the choice function exists. In response to this sort of criticism I believe that Nagel would have claimed that while the axiom of choice is apparently existential, if one considers its role in mathematics one will see that it is in reality prescriptive. He might have said, that it asserts that under certain circumstances one may infer a certain consequence. Or he might have said that according to this axiom if one *claims* that a set of certain sort exists then one may also *claim* that the choice function exists. The axiom then tells us what we may say or infer. Nagel stated, with respect to such principles that they "may be viewed as implicit definitions of the ways in which certain recurrent expressions are to be used or as consequences of other postulates for such usages."[22]

Nagel is claiming that since mathematical statements serve a prescriptive function that such statements have no ontological import. I agree that mathematical statements serve a prescriptive function such as he describes. But I deny that this implies that they have no ontological import. His conclusion, that the ontological claims are merely an "ornament", that is, not true, is unwarranted. That this is so may be seen by considering examples from natural sciences. Consider modern beliefs concerning evolution of species. If a species exists today biologists routinely infer that there was a species existing at an earlier period in the earth's history and that the species present today descended from the earlier species. Here we see modern beliefs about evolution licensing a biological inference and thus serving a prescriptive function. Statements of evolutionary theory could thus be interpreted as giving rules to be used in biological discourse. But, from the fact that the biological statements serve this prescriptive function, it does not follow that the statements have no ontological import. The statements assert the existence in the present or past of certain species. The statements in question are either true or false depending on whether the species in question exist (or existed). Indeed I would argue that there is an analogy between mathematics and biology at this point. The biological theory contains existential statements which function in part as rules. But the theory clearly has existential implications, namely, it implies that there exist species during certain time intervals in the earth's history. Similarly in mathematical theories there are existential statements such as the axiom of choice. These statements function in part as rules. But they imply the existence of choice functions in certain circumstances.

In criticizing Nagel we have only wanted to say that the acceptance of mathematical statements carries with it ontological commitments. In the paper cited above, Nagel wanted to argue that the logician had no "*a priori* insight into the most pervasive structure of things."[23] The principles of logic do not, according to Nagel, constitute knowledge "of the limiting structure of everything both actual and possible." We do not want to disagree with Nagel on this point. We do not want to imply that the mathematician has *a priori* knowledge concerning what exists or that he has knowledge of the limits of all possible reality. Philosophers such as Nagel and C.I. Lewis and Hempel and Ayer were influenced by the following argument. Truths of mathematics and logic are necessarily true. Since they are necessarily true they cannot be proved on the basis of sensory observation. Mathematical knowledge therefore must be *a priori*. Further, since they are necessarily true they must be true not merely of what exists in the actual world, but they must be true "of the limits of all possible reality." To avoid these implications regarding the mathematician's knowledge, these philosophers tried to show that mathematical principles have no ontological import. We have argued that they have not succeeded in this effort. However, the conclusion that they wished to avoid can be avoided in other ways. In particular one can deny that mathematics and logic consist of necessary truths.

In the following chapter we shall be led to reiterate some of the conclusions we have asserted in this chapter. This is necessary since in our discussion of these views we have yet to consider the work of Carnap in which they are given their most complete and forceful presentation.

24. Conclusion: In this chapter we have considered the theory that mathematical statements are analytic (as formulated by Hempel and Ayer) and the theory that mathematical statements are prescriptive (as formulated by Nagel and Lewis). We have tried to show that these theories are not successful with respect to their goal of showing that mathematical principles have no ontological implications.

NOTES

1. Hempel, (27) p. 368, and Ayer, (1) p. 79.
2. Ayer, (1) p. 79.
3. See Carnap, (10).
4. See Quine, (69) and (70).
5. See Putnam, (66).
6. The substance of this argument will not be changed if one accepts the substitutional interpretation of quantification theory. Explaining the meaning of logical principles will still presuppose a theory with ontological implications. However, the entities implied to exist will be different—expressions of a language instead of properties.
7. Ayer, (1) p. 74.
8. Hempel, (27) pp. 367–8.
9. Ayer, (1) p. 75.
10. Lewis, (47) p. 223.
11. For a definition of "queer arithmetic" see Lehman, (44).
12. The view that it is not the case that arithmetic principles should be retained in all conceivable circumstances has been criticized by Levy, (46).
13. Lewis, (47) p. 222.
14. Ayer, (1) p. 79.
15. Lewis, (47) p. 222.
16. Nagel, (57) p. 305.
17. *Ibid.*, 306.
18. Lewis, (47) p. 223.
19. Nagel, (57) p. 308.
20. *Ibid.*, p. 304.
21. For further discussion of the axiom of choice see section 25.
22. Nagel, (57) p. 319.
23. *Ibid.*, 303.

Chapter Four

Carnap's Theories

25.   As Carnap's theories concerning the nature of mathematics are rather complex, we must first explain some of his basic ideas. For this purpose we shall refer primarily to Carnap's own explanations in his monograph *Foundations of Logic and Mathematics*. Subsequently we shall consider arguments and ideas which he presented in some of his other works.

Remember that Carnap's theory is a precise and complete formulation of the idea that mathematical truths are statements which are true solely in virtue of the meanings of the words used to express the statements. In order to explain this theory we must consider the ways that words and statements of a language have meaning. According to Carnap, understanding the way that statements and terms have meaning involves consideration of what he called semantical and syntactical aspects of the language. These ideas may be explained in the following way: If one studies a language one may consider such factors as what causes certain acts of speech (or of writing), what are the effects of acts of speech on those who hear them, what social function certain acts of speech play in a society. The study of these factors has been called pragmatics. The study of pragmatics clearly involves consideration of the acts of language users, the context or environment in which language is used, the effects of uses of language on other people in the environment, etc. Through observing language users one may acquire knowledge of such matters with respect to the language under observation.

Such observation will also lead to knowledge of the following sorts: One may learn, for example, that certain signs of the language designate certain objects and one may learn that other signs of the language are used to refer to certain qualities of objects or to relationships in which objects stand to each other. Thus, one may learn by observation according to Carnap that in English the sign 'moon' is used to refer to the moon. And one can learn that expressions of the form "... is white" are used to designate the color of such objects as the moon, pieces of paper, skin of some people, etc. That study of language which consists in considering only the relationships between signs and what they designate is called "semantics." Finally, in studying a language one can observe, according to Carnap, the kinds of objects that serve as signs in the language and one can also observe the order or sequences of these kinds of signs as used. For example, one might observe that certain signs of the English language consist of the form "...... is ------" in which words of one form occur in the first blank and words of a second form occur in the secon blank. The study of the forms of signs and of their order is called syntax or logical syntax.

In Carnap's theory the idea that mathematical truths are statements which are true in virtue of the meanings of their terms is understood in the following way: mathematical truths are statements which are true in virtue of semantical and syntactical rules of the language. That is to say, mathematical truths are statements whose truth is determined solely by rules which relate to (1) kinds of signs occurring in the statement and their serial order and (2) relations between signs and the objects, properties or relations designated by those signs. Other aspects of language studied in pragmatics do not enter into the meaning of the signs. For example, factors which cause the use of a sentence or functions the sentence serves in the society are not in general part of its meaning. (It may happen, of course, that the object designated by some signs of a sentence are part of the cause of the utterance of that sentence.)

To verify Carnap's claim concerning the nature of mathematical truths it would appear necessary to state the semantical and syntactical rules of English or of some other language such as German. However, Carnap believed that this was an impractical task. Languages such as these are too unsystematic to allow for simple and orderly presentation of their syntactical and semantical rules. To overcome this difficulty Carnap was led to construct what he called a "semantical system" and a "syntactical system." These systems were to correspond to the semantical and syntactical parts of natural languages. Sentences which could be expressed in the natural languages (at least mathematical sentences) could also be expressed in such a semantical system. In such a system the rules of syntax and semantics would however be precise and orderly.

To show how Carnap's theory concerning mathematical truth was to be verified we shall in the following sections present first a syntactical system and then certain semantical rules for the system thus obtaining a semantical system. The resulting semantical system was thought, by Carnap, to correspond to our ordinary rules of logic. We shall then show how certain mathematical sentences can be formulated in the semantical system. Carnap's basic theory is that these mathematical sentences correspond to truths of mathematics (in natural languages) and that the truth of these sentences is determined by the semantical rules of the semantical system. We shall attempt to keep our presentation sufficiently simple so that it can be followed by people who are not mathematicians or mathematical logicians. Only after we have fully presented Carnap's theory will we engage in critical discussion of it with reference to the basic questions of this book.

26.   The syntactic system which we shall present consists of certain signs called *logical* signs. There are several kinds of such signs. These are:

1. Sentential variables: $p, q, p_1, q_1 \ldots$
2. Individual variables: $x, y, x_1, y_1 x_2, y_2, \ldots$
3. Predicate variables: $F, G, F_1, F_2, \ldots$
4. Grouping indicators: (,), [,].
5. Constants: $\supset$, $\sim$.
6. Quantifiers: $\exists$, $\forall$.

We shall refer to sequences of such signs as formulas. In this syntactical system only certain sequences of signs may occur. These are given by the formation rules and the permitted formulas are called *well-formed formulas*. Some of the formation rules of the system are:

1. Any predicate variable followed by any number of individual or predicate variables is a well-formed formula as is any sentential variable.
2. If P and Q are well-formed formulas then so is $(P \supset Q)$.
3. If P is a well-formed formula then so is $(\sim P)$.
4. If P is a well-formed formula then so is $(\forall \quad)P$. (Where the blank space after the '$\forall$' can be occupied by any variable.)

Examples of some well-formed formulae are: Fx, FF, $(\forall x)Fx$, $(\forall F)Fx$, $(Fx \supset Gx)$, etc.

Ordinary rules of logic enable us to deduce some sentences from other sentences. The sentences in a deduction form a sequence. In a syntactical system which is to correspond to our logic there may be sequences of well-formed formulae. These sequences may be obtained by stipulating that certain formulae are to be given as primitive formulae and then stating the rules by which sequences of formulae may be generated from the primitive formulae. The rules which allow for the generation of sequences are called transformation rules. Some of the primitive formulae of the system are

1. $[(\sim p) \supset p] \supset p$
2. $p \supset [(\sim p) \supset q]$
3. $(\forall x) Fx \supset Fy$
4. $(\forall G)(\forall F)(\forall y)[(\forall x)(Gy \supset Fx) \supset (Gy \supset (\forall x) Fx)]$

Some of the transformation rules for the system are

1. From two formulas of the form P and $(P \supset Q)$ one may obtain the formula Q.
2. From a formula of the form $(A \supset Bx)$ in which the part after the ' $\supset$ ' contains the variable 'x' while 'x' does not occur in A, the formula $(A \supset (\forall x) Bx)$ may be obtained.

The specification of the syntactical system which we have given serves primarily at this point as an example of the sort of thing that Carnap had in mind in his reference to a syntactical system. In order that the formulae of such a system be made to correspond to valid principles of ordinary logic semantical rules must be given. In the absence of semantical rules the above formulae are meaningless. The addition of the semantical rules turns the syntactical system into a semantical system.

In *Foundations of Logic and Mathematics* Carnap spoke of two sorts of semantical rules. Rules of the first sort assign designata as the referents of term. In the syntactical system we have specified above there may be well-formed formulae such as "Fx" and "GF." Semantical rules for these formulae of the kind indicated by Carnap are:

1. 'x' designates this book.
2. 'F' designates the property of being long.
3. 'G' designates the property of being widely manifest.

Given these three semantical rules, "Fx" becomes "This book is long", and "GF" becomes "The property of being long is widely manifest."

In addition to semantical rules which assign designata to terms there are semantical rules which specify the conditions under which statements are true. Thus we might have

1. A sentence of the form "*Fx*" is true if and only if the object designated by '*x*' has the property designated by '*F*'.
2. A sentence of the form "∼*Fx*" is true if and only if the sentence "*Fx*" is not true.
3. A sentence of the form $P \supset Q$ is true if and only if either *P* is not true or *Q* is true.
4. A sentence which is not true is false.
5. A sentence of the form "$(\forall x)Fx$" is true if and only if "*Fx*" is true for every individual, i.e., for every individual in the universe if '*x*' is assigned that individual as designatum then "*Fx*" is true. (This would be true only if every individual had the property designated by '*F*'.)[1]

In presenting the above example of a semantical system we have not attempted to present a system of logic which is complete with respect to validity, thus, there will certainly be principles of logic which are valid which do not correspond to formulas that can be deduced from the axioms we have stated by the rules that we have stated. Those readers interested in formulation of complete systems of first-order quantificational logic may consult any one of a range of logic texts.

A distinction which Carnap believed was quite important involves noting that it is possible that the truth of some sentences is determined solely by the way that the semantical rules apply to the sentence. The semantical rules guarantee that the sentence is true. In other cases the semantical rules guarantee that the sentence is false. The semantical rules we have stated guarantee that the following sentence is true (no matter what predicate is designated by '*F*'):

$$(\forall x)\ (Fx \supset Fx)$$

and they guarantee that the following sentence is false:

$$\sim(\forall x)[Fx \supset Fx]$$

If the semantical rules of a semantical system are sufficient to guarantee that some sentence *P* of the system is true, then Carnap called the sentence *L-true*. If the semantical rules guarantee that a sentence *P* is false then Carnap called the sentence *L-False*. Sentences which are either L-True or L-False are called *L-determinate*. Sentences which are not L-determinate were called by Carnap *factual*.

27.   Since we are concerned with questions of ontology in relation to mathematical theories we must consider now the language of a mathematical theory. In Carnap's view, underlying such a theory there will be a syntactical system (also called a calculus). Then, of course, there will be semantical rules interpreting the signs of the calculus. If one were just making a calculus for no purpose at all then one could form it in any way he chose. However, since we want our calculus to be the calculus of a mathematical theory, this leads to the formation of certain sorts of calculi rather than others. It does not, however, determine a calculus uniquely. Carnap believed that for mathematical theories the underlying calculus includes as a part a calculus suitable for logic. However, it might appear that additional initial formulae and additional semantical rules might be needed in order to extend the logical calculus to one suitable for mathematics. However, according to Carnap, such additions to the semantical system of logic are not necessary. He believed that the statements of mathematics could be translated into statements of logic, that is, into statements which correspond to formulae of the basic logical calculus as interpreted by the semantical rules of logic. Of course, Carnap thought that the truths of mathematics are, when properly translated into statements of logic, L-true. He attempted, in the *Foundations of Logic and Mathematics*, to show that elementary arithmetic, based on Peano's postulates, consists of a system of L-true statements. Let us now consider how this may be done.[2]

Carnap presented Peano's postulates as follows:

P1. b is an N.

P2. For every x, if x is an N, then x' is an N.

P3. For every x,y, if (x is an N and y is an N and $x'=y'$) then $x=y$.

P4. For every x, if x is an N, then it is not the case that $b=x'$.

P5. For every F, if (b is an F and, for every x (if x is an F then x' is an F)) then (for every y, if y is an N then y is an F).

Carnap regarded these postulates as being part of a syntactical system or calculus which contains the signs 'b', 'N' and '$'$'. This calculus would be turned into the semantical system of arithmetic as normally understood by the adoption of the following semantical rules (according to Carnap).

1. '*b*' designates the number 0.
2. '*N*' designates the class of finite cardinal numbers.
3. If "..." designates a cardinal number then "...′" designates the next one (the immediate successor).

Given these semantical rules, Peano's postulates become the following true arithmetical statements:

*P1. 0 if a finite cardinal number.
*P2. For every *x*, if *x* is a finite cardinal number then the immediate successor of *x* is a finite cardinal number.
*P3. For every *x* and for every *y* (if *x* and *y* are finite cardinal numbers such that the immediate successor of *x*=the immediate successor of *y*) then *x*= *y*.
*P4. For every *x*, if *x* is a finite cardinal number then 0 is not the immediate successor of *x*.
*P5. For every *F*, if (0 is an *F* and, for every *x* (if *x* is an *F* then the immediate successor of *x* is an *F*)) then (for every *y*, if *y* is a finite cardinal number then *y* is an *F*).

In order to develop Carnap's idea, it is now necessary to show that *P1–*P5 are L-true statements of logic that is of the (completed version) semantical system which we have described above (or of some alternative equivalent system of logic). However, in that semantical system we did not have expressions '0', "finite cardinal number" or "immediate successor." How then can the Peano postulates (as interpreted by the semantical rules above) be shown to be L-true statements of logic? The answer to this question, Carnap believed, involved showing that expressions having the same meaning as the arithmetical expressions '0', "successor" and "finite cardinal number" can be introduced into the semantical system of logic through the use of certain definitions. Through the use of these definitions the statements *P1–*P5 can be shown to be merely abbreviations for L-true statements of logic. Some of these definitions as Carnap gives them are as follows:[3]

> "*x*= *y*" for "for every (property) *F*, if *x* is an *F* then *y* is an *F*." Analogously for any higher level.
> "*F* is an 0" for "not (for some *x*, *x* is an *F*)."
> "*F* is an *m*+" for "for some *G*, for some *x*, for every *y* [(*y* is an *F* if and only if (*y* is a *G* or *y*= *x*)) and *G* is an *m* and, not *x* is a *G*]."
> "*m* is a finite cardinal number" for "for every (property

of numbers) K, if (0 is a K and, for every n (if n is a K then $n+$ is a K)) then m is a K."

Two brief explanations are necessary at this point with regard to these definitions. First, Carnap is using the sign '+' to signify the successor function, that is to say, '$m+$' signifies the successor of 'm'. Further the sign 'm' is being used to refer to the appropriate predicate variables of the semantical system.

Secondly, in the statement in which identity is defined there is the expression "Analogously for any higher level." Here Carnap is assuming that the syntactical system is somewhat more complex than that we have given above (section 26). There we distinguished between individual variables and predicate variables. However, for reasons having to do with avoiding logical inconsistency, Carnap is using a finer distinction.[4] The variables are assumed divided into an indefinite number of levels. Individual variables form the zero level. Then there are variables for level 1, variables for level 2, variables for level 3, etc. The syntactical system, to which Carnap is referring, has more complex formation rules for formulae also. The rules involve restrictions with respect to level in the formation of formulae by sequences of variables. The above definition of the sign '=' defines the sign only for individual variables. Analogous definitions are necessary to define the sign for use with higher level variables.

We should note at this point that postulate *P5 involves higher level variables and that in indicating the semantical rules above (section 26) we did not give a semantical rule for formulae involving such variables. The semantical rule which Carnap gives for such formula is the following:[5]

5#. A sentence of the form "for every ...,---" is true if and only if all entities belonging to the range of the variable "..." have the property designated by "---" with respect to "...."

The entities in the range of individual variables are individuals. The entities in the range of variables of level 1 are properties of individuals. The entities in the range of variables of the next higher level are properties of properties of individuals, etc. With the adoption of semantical rule 5# we can dispense with semantical rule 5 given earlier.

Before concluding this section let me briefly explain the above definitions in informal terms. To say that x is identical with y is to say that for any property F, if x has the property F then y has the property F. To say that F is a zero is to say that no object has the property F. To say that F is a successor of m, i.e., F is an $m+$, is to say that there is some property G which has the property m and there is some x which lacks the property G and no matter what object you pick, it has the property F if and only if either it has the property G or is identical with x. To say that m is a finite cardinal number is to say that m has every property which belongs to 0 and all the successors of 0.

28.   It is not difficult to convince ourselves that the statements corresponding to Peano's postulates which can be formulated in the semantical system of ordinary logic are indeed L-true. By considering some of the postulates we can easily see that the truth of these postulates follows from the semantical rules. Let us briefly show that this is so. Consider first Peano's first postulate namely

> 0 is a finite cardinal number

If we substitute the definition of the expression "m is a finite cardinal number" and put 0 in place of 'm' the result is (ignoring type restrictions)

> For every K, [if (0 is a K and, for every n (if n is a K then $n+$ is a K)) then 0 is a K]

Clearly this statement is L-true since every instance of

> if [0 is a K and for every n (if n is a K then $n+$ is a K) then 0 is a K]

is true in virtue of the semantical rule governing statements of the form $P \supset Q$.

Again, consider the second postulate, namely

> For every m, if m is a finite cardinal number, then $m+$ is a finite cardinal number.

This will be true providing there is no m such that m is a finite cardinal number but $m+$ is not a finite cardinal number. Can we show that there is no such m through considering the semantical rules governing the logical signs? We cannot answer this by reference to the semantical rules we have already stated because we did not above state the semantical rule for conjunction. But this rule is quite obvious. It is as follows:

A statement of the form P and Q is true if and only if both P and Q are true.

Now, is every instance of

S) $(\forall m) [(\forall K)((K_0$ and $(\forall n) (K_n \supset K_n+)) \supset K_m] \supset [(\forall K)(K_0$ and $(\forall_n) (K_n+ K_n+)) \supset K_m+)]$

true. (Clearly the antecedent and the consequent in statement S are "m is a finite cardinal number" and "m+ is a finite cardinal number." In virtue of the semantical rule for the universal quantifier, S will be true if the conditional sentence within the scope of "$(\forall m)$" is true for every object (of suitable type) in the domain. So consider any such object. The conditional in question will be true if either its antecedent is false or its consequent is true. Since the semantical rules stated above imply that every statement is either true or false, we may argue as follows: If for the object in the domain that we have picked the antecedent of the conditional is false then the conditional is true. But, if for the object we have picked the antecedent is true then we have for the object m in question that

A) $(\forall K)[(K_0$ and $(\forall n) (K_n \supset K_n+)) \supset K_m)$

is true. We have to show for the m in question that

B) $(\forall K)((K_0$ and $(\forall n) (K_n \supset K_n+) \supset K_m+)$

is true. But (A) is true if and only if either its antecedent is false or its consequent is true. If the antecedent of (A) is false then (B) is true (since (B) has the same antecedent as (A).) But if the antecedent and consequent of (A) are true when we know that K_m is true and (in virtue of the semantical rule for conjunction) that $(\forall n) (K_n \supset K_{n+})$ is also true. But then, in virtue of the semantical rule for the universal quantifier "(n)" we know that "$K_m \supset K_{m+}$" is true and since K_m is true so is K_{m+}. But this shows that (B) is true since its consequent is true. Since we have shown that (B) is true whenever (A) is we have shown that (S) is true and we have done this solely by reference to semantical rules. In similar manner we could show that each of the other postulates is true in virtue of the semantical rules. Can we conclude that Carnap has provided an argument which shows that mathematical statements such as Peano's postulates are true in virtue of the meanings of their terms and thus that such mathematical statements have no ontological import? I shall try to show that Carnap has not accomplished that much.

First, we may ask, whether the statements which Carnap has shown to be L-true are really Peano's postulates. Consider Peano's first postulate. This is normally expressed as the statement "0 is a number." In this statement the expression '0' is normally understood to designate the number 0 and the expression "number" is understood to designate the counting numbers (or the natural numbers). This postulate appears to claim that some object in the universe is a member of a particular set of objects. This is a statement about one particular object. But, under Carnap's translation, this postulate is no longer a statement about one particular object. It is rather a universal generalization. The first postulate as formulated by Carnap says that any property K is possessed by zero and such that if it is possessed by an object then it is also possessed by the successor of the object, is possessed by 0. We may indeed wonder whether Carnap has shown that *Peano's postulates* are L-true.

Carnap might have responded to the above objection by claiming that it is not the case that '0' is normally understood as designating a particular object. Carnap would certainly have thought this since the number zero is not an observable object. The way of understanding the first postulate, which I have claimed is the normal way of understanding it, would, according to Carnap, be meaningless and so could not be the normal way of understanding it. This reply is not satisfactory however. It rests ultimately on the so-called "verification theory of meaning" which has been rejected by philosophers for sufficiently good reasons. Discussion of these reasons would be out of place here.[6]

Let us turn to a second criticism. In order to substantiate the position that statements of mathematics have no ontological import in the manner developed by Carnap, it would be necessary that the axioms of the semantical system of logic, on which Carnap's argument rests, have no ontological import. But, on the contrary, it appears that the axioms of the semantical system in question have ontological import. For one thing, even the axioms given above (which are similar to axioms which Carnap states) appear to presuppose the existence of objects and properties. The natural way of interpreting axioms of higher-order logic is as presupposing a (non-empty) universe in which there exist individuals and properties and properties of properties, etc. The acceptance of these logical principles appears to involve the acceptance of certain ontological assumptions. (Even if one rejects the natural way of interpreting the axioms of higher-order logic and claims instead that such axioms are to be interpreted with respect to the expressions which may be substituted for the variables of the axioms, the claim that the system of logic involves ontological assumptions is not avoided.)

For another thing, the full semantical system of logic which Carnap must presuppose in order to justify his claim to be able to construct all of classical mathematics within this semantical system, contains axioms which are explicitly existential. Informally it is not difficult to see this. Peano's postulates imply that there is no last counting number. Given that the numbers do not "circle back on themselves", this implies that there are an infinite number of counting numbers. The axioms of the semantical system thus must imply the existence of this infinite collection. Indeed, since we may assume that Carnap was utilizing the system of logic developed by Russell and Whitehead, namely *Principia Mathematica*, for his complete semantical system, Carnap was assuming explicitly an axiom of infinity. This implies that there are an infinite number of objects (or of properties). Clearly this is an ontological assumption.

A further axiom which was adopted by Russell and Whitehead is the axiom of choice (which they called the multiplicative axiom). In section 21 we stated this axiom as follows:

> In any set S of disjoint sets not containing the null set, there exists a function f whose domain is S and such that for each member m of S, $f(m)$ is a member of m.

Let us briefly explain the content of this axiom. To say that a function exists is to say that there is a set of objects which are arguments of the function and that there is a set of objects which are values of the function. In the case of the choice function, the set of arguments are the sets which are members of S and the set of values is the set containing the individual members of members of S which are picked out by the choice function. But this axiom, like the axiom of infinity, is an existential assumption. This axiom could be expressed in the symbolism of the syntactical system we have given above. But the semantical rules we have stated (essentially the semantical rules used by Carnap) do not imply that the statement is true.

To convince ourselves (informally) that the axiom of infinity is not true simply in virtue of semantical rules which give the customary meanings of the logical connectives and the universal quantifier, it is only necessary to consider the supposition that the number of entities in the universe (individuals and properties and properties of properties) is finite. The supposition that the number of entities in the universe is finite is, of course, the supposition that there is some counting number n such that each entity in the universe could be paired with a number less than or equal to n so that no number was paired with more than one entity and no entity was paired with more than one number. In other words, this is the supposition that there is a one-to-one correspondence from the entities in the universe onto the counting numbers up to n. We may suppose that this supposition is true. We are not prevented from doing this by the semantical rules stated above. If the axiom of infinity were true in virtue of semantical rules, then the semantical rules would rule out (be inconsistent with) the supposition that the number of entities in the universe is finite.[7]

In similar manner we can argue that the axiom of choice is not true in virtue of the semantical rules. If it were true in virtue of the semantical rules then the semantical rules would be inconsistent with describing a set of sets in which the axiom of choice is false. It is not difficult to describe such a set of sets and I see no reason for thinking that such a set is inconsistent with the semantical rules. A set of sets which cannot exist if the axiom of choice is true is the following: The members of the set S are a set containing the numbers one and two and a set containing the numbers 3 and 4. S has two sets as members in this case and the members are disjoint and the null set is not a member of S. But the set of values of the choice function is not a member of S either.

Carnap, in trying to show that mathematical principles were L-true was clearly trying to show that mathematical principles have no ontological import. We have criticized Carnap's position. We have suggested that his translations of Peano's postulates into the semantical system he sketches may not render the customary meaning of the postulates. Further, while Carnap does not give a complete semantical system within which principles of classical mathematics such as the axioms of real number theory could be expressed, we have argued that in the system which he had in mind, namely *Principia Mathematica* there are axioms which are not L-true and which do have ontological implications. Thus, one cannot appeal to this semantical system to try to show that mathematical principles have no ontological import. We shall continue this criticism in the next section where we consider the problems posed by the use of "impredicative" definitions in mathematics and by the occurrence in *Principia Mathematica* of the axiom of reducibility.

29.   Mathmetics includes not only arithmetic as based on the Peano postulates but also real number theory as based on axioms such as those given in section 4 as well as many other branches based on still other axioms. Carnap hoped to be able to show that mathematical principles, in whatever area, were L-true by showing that such principles could be translated into statements of a semantical system such as the above and which would then be L-true in accord with the rules of the semantical system. We have argued that in order to make the semantical system strong enough to accomplish the intended purpose it would have to take as first principles (or alternatively as transformation rules) principles which are not L-true, e.g., the axioms of infinity and choice. Another axiom of *Principia Mathmetica* which we shall have to consider is the axiom of reducibility. This axiom plays an essential role in Russell and Whitehead's treatment of real number theory and I want now to discuss that role.

   To motivate this discussion let us call to mind the fact noted in section 4 that the rational numbers form an ordered field which is not complete. In pictorial terms, we might say that the rational number sequence has gaps in it. If our number theory underlying the techniques we use for solving equations, i.e., finding their roots were based on the rational numbers rather than the real numbers then we should have to say that certain equations, which we now think of as having roots, could not be solved, e.g., the equation $x^2 = 2$. Problems represented by such equations would have no precise solution. For example, the problem of finding the length of the diagonal of the square whose side has length one (the unit square). This might not bother engineers (and others who use mathematics) in physical constructions. They could do what they have to do anyway in cases where some physical magnitude is represented by an irrational number. That is, they could approximate the physical magnitude with some rational numbers. The inconvenience in other scientific problems would probably be somewhat greater. At any rate, some inferential processes, for example in calculus, are justified mathematically by reference to assumptions that are not satisfied by the set of rational numbers due to the above noted gaps. For example, in applications of mathematics in certain physical problems it is often necessary to assume that a certain function is differentiable over some segment or region. This assumption presupposes that there are no gaps in the segment. Even if approximate solutions to the problem can be worked out without differentiating, the inconvenience of doing this has led scientists to place great reliance of the calculus and thereby on the real number theory which underlies it.

As noted earlier, the rational numbers are an ordered field. The simplest way to develop the theory of real numbers in relation to the rational numbers is to assume that the gaps are filled. This is done by the assumption that every set of real numbers which is non-empty and has an upper bound has a least upper bound as well. But consider now the definition of a least upper bound: the least upper bound of a set of real numbers is an upper bound which is less than all of the remaining upper bounds of the set, that is, it is the least of the upper bounds. This sort of definition violates a principle which Russell believed must not be violated if contradictions such as the liar paradox and Russell's paradox were to be avoided. Russell referred to this principle as the "vicious circle principle." It has been much discussed of late.[8] Russell provided several formulations of it. For example he said "Whatever involves all of a collection must not be one of the collection; or, conversely: If, provided a certain collection had a total, it would have members only definable in terms of that total, then the said collection has no total."[9] The specification of the least upper bound of a set of real numbers involves all of the collection of upper bounds, i.e., it is defined as the least of the set of all upper bounds of the collection. In this definition one assumes that a certain set has a total, namely the set of upper bounds of the collection, and then one defines a specific element of this set by reference to the total that is, one specifies the least one of the set. This definition apparently violates the vicious circle principle. The definition of "m is a finite cardinal number" given above also appears to violate the vicious circle principle. In that definition the property of "being a finite cardinal number" is defined. But the definition refers to all properties. Thus, a member of the collection of properties is defined by reference to the collection as a whole. Defintions which violate the vicious circle principle were called by Russell "impredicative definitions."

The mathematician Henri Poincaré agreed with Russell that definitions in mathematics must not violate the vicious circle principle. He also referred to definitions which violate this principle as "impredicative." He attempted to explain the fallaciousness of statements which violate this principle. According to Poincaré, when one gives a definition of a term, one is presupposing a fixed classification of objects, that is, a classification of objects that is settled and with reference to which the term may be defined. For example, all objects should be divisible into a class having certain properties and into one lacking those properties and then a new term may be defined with reference to one or another of these classes. The trouble with the definition of "is a finite cardinal number" and other impredicative definitions, according to Poincaré, is that the classes are not fixed prior to the definition. He regarded such a definition as specifying a new entity or property belonging to one of the classes of objects with reference to which the definition was given but not contained in that class prior to the creation of the definition. The creation of the definition thus disorders the classification—unfixes the classification—which was utilized to give the definition. Since the terms used to give the definition have no fixed meaning, that is, since their meaning is changed by the definition itself, the defined term gets no clear meaning. Consequently inferences in mathematics using valid principles of logic are not really valid and contradictions result. The fallacy involved when one uses impredicative definitions seems, in Poincaré's view, to be ultimately a fallacy of ambiguity.[10]

In order to conform to the vicious circle principle in his logical theory, Russell introduced the theory of types and divided propositions and propositional functions into various *orders*. The theory he devised is quite complex and is now referred to as the ramified theory of types.[11] I shall try to explain enough of the theory so that the reader may be able to understand the axiom of reducibility.

Consider a proposition such as "John is tall." This proposition contains the term or constant "John" as well as the constant "is tall." If we replace the constant "John" with 'x' the resulting expression "x is tall" is called a *propositional function*. Russell referred to expressions such as the 'x' in propositional functions as *real variables*. He thought of the 'x' as referring ambiguously to any one of a number of entities within a certain range or domain. The 'x' varies over the range in question. Russell thought that in mathematics one asserted propositional functions such as "x is tall" and that such assertions were indeed necessary for deductive inferences. However, he held that assertion of a propositional function should not be confused with the assertion that a certain propositional function is true for all values of the variable in the range in question. The assertion that "x is tall" is true for all values of the propositional function would have been represented by Russell as "$(\forall x)$ (x is tall)." According to Russell this is a definite assertion about every member of a certain range and not an expression containing a term which refers ambiguously. However, "$(\forall x)$ (x is tall)" also contains 'x' but, since the 'x' is not, in Russell's view, serving as an ambiguous referring expression, he called the 'x' an *apparent variable*. "$(\forall x)$ (x is tall)" is a proposition; it is not a propositional function.

In the proposition "John is tall" the term "John" refers to one individual. In the proposition "$(\forall x)$ (x is tall)" the apparent variable is used to refer to all individuals of a certain range. A proposition of either of these sorts, that is a proposition which contains no apparent variables or one which contains apparent variables but which refers only to individuals, is called a *first-order proposition*. Russell believed that one could make assertions about first-order propositions. Propositions in which the apparent variable refers to first-order propositions are called second-order propositions. Clearly, we can envision a hierarchy of propositions. But, for the development of logic, Russell believed it more useful to develop a hierarchy of propositional functions. Suppose that we have a propositional function 'ϕ' the argument of which is always an individual as in expressions such as "ϕ John." Here 'ϕ' ranges over properties of individuals such as John. Russell would have thought of it as referring ambiguously to such properties. If 'ϕ' is replaced in the expression "ϕ John" by the name of a definite property the result is a first-order proposition. The resulting first-order proposition is said to be a value of this propositional function. Now, Russell said that "a function whose argument is an individual and whose value is always a first-order proposition will be called a *first-order function*. A function involving a first-order function or proposition as apparent variable will be called a *second-order function*, and so on."[12] For example, if 'ϕ' is a first-order function then "$(\forall \phi) \psi \phi$" is a second-order function. Now, "A function of one variable which is of the order next above that of its argument will be called a *predicative function*; the same name will be given to a function of several variables if there is one among these variables in respect of which the function becomes predicative when values are assigned to all the other variables."[13]

Russell defined the term *type* as being the set of objects which can be arguments of a propositional function. For example with respect to the propositional function "x is tall", John is one of the arguments: with respect to the propositional function "X is a short sentence", "John is tall" is one of the arguments. Since "John is tall" and "John" are not arguments within the range of the same propositional function they are not of the same type.

The logical theory devised by Russell in order to avoid the paradoxes such as Russell's paradox, the paradox of the liar, Richard's paradox, etc., included the restriction that propositional functions be predicative functions.[14] Russell argued that this restriction was sufficient for avoiding the paradoxes. However, he realized too that definitions of important concepts in mathematics, e.g., the least upper bound of a set, involved impredicative functions. He believed that the problem of formulating definitions adequate for the mathematical concepts in question, could be solved by the adoption of the axiom of reducibility. He stated this assumption as follows: "Every function is equivalent, for all its values, to some predicative function of the same argument."[15] The adoption of the axiom of reducibility permitted the assumption that there is a predicative function equivalent to the function which enters into the definition of the "least upper bound." Similarly, other mathematical principles involving the use of impredicative functions could be used. However, Russell believed, that since the theory of types prevented reference to all objects or to all properties, that paradoxes would not be reintroduced by the adoption of the axiom of reducibility.

There seems to be no question that the axiom of reducibility is not L-true and Carnap did not consider it as such. It is clearly an existential axiom in that it affirms the existence of certain entities, namely, predicative functions. Rather than trying to argue that the axiom of reducibility was L-true, Carnap argued that it could be eliminated as a fundamental principle of the semantical system. Here, Carnap was following the work of Ramsey. Ramsey had argued that the paradoxes could be divided into two groups.[16] The paradoxes of one group, e.g., Russell's paradox, which are now called logical paradoxes, could be eliminated from the semantical system by the adoption of a simplified version of the theory of types (now called the simple theory of types). This simplified version did not conform to the vicious circle principle; that is, impredicative definitions are allowed. Further, Ramsey argued, that the other paradoxes, such as Richard's paradox, arise due to the occurrence of such expressions as "is nameable" or "is describable." However, these expressions do not need to occur at all in a logical theory and thus they could be avoided in logic without having to adhere to the vicious circle principle. Carnap agreed with Ramsey that impredicative definitions could be allowed in conjunction with the simple theory of types. Thus, even though the axiom of reducibility is not L-true, Carnap did not regard this as undermining his view of logic and mathematics since he could hold that this principle is not a logical or mathematical principle which he must accept.[17]

However, the logical system which, we are assuming, Carnap adopted, still contains the axioms of infinity and choice. And we have argued that these are not L-true. How might Carnap have responded to this objection? Presumably he would have followed Ramsey at this point also. In the above cited work, Ramsey claimed that these axioms are tautologies (if properly understood). However, Ramsey's arguments for his view that the axioms of choice and of infinity are tautologies seem to me to be mistaken. Indeed in the case of the axiom of choice (which he, following Russell, called the multiplicative axiom) there is hardly any argument at all. Ramsey said that the multiplicative axiom

> asserts that given any existent class K of existent classes, there is a class having exactly one member in common with each member of K. If by "class" we mean ... any set homogenous in type not necessarily definable by a function which is not merely a function in extension, the multiplicative axiom seems to me the most evident tautology.[18]

This statement is hardly persuasive especially as we have tried to show how one may describe a set of sets in which the axiom fails above.

With regard to the axiom of infinity, so far as I understand Ramsey's argument, it seems to be that either the axiom is meaningful and tautologous or else it is meaningless. It is not meaningless and so it is a tautology. But, we clearly do not have to accept the first premise of the argument. The proposition that the universe is finite appears both meaningful and consistent and if this is correct then the first premise of Ramsey's argument is mistaken. In any case, even if Carnap were inclined to agree with Ramsey in regard to these two axioms, I do not see how he could justify the claim that the axioms are L-true with reference to the semantical rules that he has given.

A defender of Carnap at this point might raise the question of the existence of a syntactical system based on some logical theory other than that of *Principia Mathematica*. Is there an alternative logical theory which is adequate to justify Carnap's view that truths of classical mathematics are L-true and in which the axioms of logic are L-true? The answer to this is, I believe, negative. Many alternative logics have been proposed.[19] But, all such logics to which Carnap might have appealed contain axioms which are existential and so would not be L-true.

One final effort at defending Carnap's view at this point might involve the proposal of additional "semantical rules." If there were such semantical rules then perhaps by appeal to such rules we could show that the axioms of infinity and choice were indeed L-true. To this suggestion I suppose that I should reply by taking a wait-and-see attitude. Perhaps there are such rules. After all, Ramsey at least thought that these axioms are tautologies. However, I rather doubt that I would agree that there are any rules that might be proposed as "semantical rules" for logic and which had the result that the axioms of choice and infinity are L-true, but which could be used to show that these principles had no ontological import. After all, these principles assert that sets satisfying certain conditions exist.

30. In his paper "Empiricism, Semantics, and Ontology"[20] Carnap has provided arguments against the type of criticisms that we have been making. We have been claiming that he has not succeeded in showing that mathematics is free of ontological commitments. He has provided no grounds for denying that mathematical entities exist or are real. On the contrary, we have argued, the acceptance of mathematical statements as true implies a commitment to the existence of mathematical entities. In the paper in question Carnap argues that the acceptance of a mathematical theory, or of any scientific theory for that matter, does not involve ontological commitments to any entities—queer or not. Let us now turn to a consideration of this view and of the arguments which are offered in its support.

The argument of Carnap's paper rests on a distinction which Carnap believed is well-grounded. According to Carnap, if someone wishes to speak of a certain kind of entity, he must have available a suitable "linguistic framework." Of course, we can understand this term as referring to a syntactical system to which have been added suitable semantical rules. In order to discuss a kind of entity, there must be in the framework general terms for making statements about the properties or relations of the entities in question. These could be obtained through having in the syntactical system certain constant terms which could be interpreted as designating the required properties or relations. The semantical rules supply the interpretation, i.e., they correlate the terms with the properties, relations or other objects. Carnap also suggests that in a linguistic framework suitable for discussing several kinds of entities there should be variables of distinctive types for referring to the kinds of entities in question. But, it is not clear to me why advantages would be gained by the adoption of different styles of variables and this does not seem to be essential to his view. Quine also has discussed this aspect of Carnap's view and finds the reference to style of variables inessential for Carnap's position.[21] (Some gain might be made with respect to ease of recognition of expressions of the language.)

With respect to such frameworks, Carnap claims that one can distinguish two kinds of questions with respect to the existence of entities. He calls these internal questions and external questions. He says that the internal questions are "questions of the existence of certain entities of the new kind within the framework." "Questions concerning the existence of reality of the system of entities as a whole" are external questions.[22] I take it that what he had in mind by this distinction is something along the following lines: An internal question of existence concerns a kind of entity which can be referred to and described by terms of the linguistic framework. Further, it is a question for which an answer may be expected through the use of methods spelled out within the linguistic framework. For example, in the example of a syntactical system sketched out above under the usual interpretation of the logical signs, we might expect that certain statements could be deduced from the first principles of the system. An internal question with respect to this system would be one for which we might expect to achieve an answer through deducing an appropriate theorem from the first principles. Carnap referred to frameworks of that sort as logical frameworks. In other frameworks he suggested the method used to answer questions within the framework would involve making observations. Such frameworks he called factual frameworks.

According to Carnap, within a framework for real-number theory we could ask such questions as whether there are prime numbers, whether there is largest prime number, whether there is a largest pair of twin primes, etc. However, someone might also ask, what about the whole system of real numbers? Is the system as a whole real? This is an external question of existence and, according to Carnap, is a pseudo-question. He affirms that any assertion concerning the reality of the system of entities of a framework taken as a whole "is a pseudo-statement without cognitive content."[23] There are, of course, other external questions such as the question of whether to accept the linguistic framework. But, according to Carnap, this is a practical question rather than a theoretical one. "The acceptance cannot be judged as being either true or false because it is not an assertion. It can only be judged as being more or less expedient, fruitful. ..."[24] The "acceptance of a linguistic framework must not be regarded as implying a metaphysical doctrine concerning the reality of the entities in question."[25]

The reader will no doubt have noticed that in describing the notion of an "internal question" I did not say that these are questions for which the methods spelled out in the framework are sufficient for giving an answer. For example, in a "logical" or "mathematical" framework question may be raised as to whether such and such a statement is a theorem. It is entirely possible that the methods contained in a framework for answering this sort of question will be incapable of giving a yes or no answer in some particular case. This is even more clearly true with respect to factual frameworks. For example, within a framework for evolutionary biology we may not be able to decide whether two bone specimens belonged to individuals of two different species or whether they belonged to individuals of one species. Further, one framework may be preferable to another just in point of being able to provide solutions to questions for which the other could provide no solutions.

Carnap allows that frameworks may be accepted or rejected. Presumably then there may be criteria or standards of evaluation with reference to which a decision among alternative frameworks can be made. Carnap, points out, for example, that as a consequence of observations we find describing spatial relations by using three-dimensional coordinate systems more satisfactory than either two- or four-dimensional systems. He also suggests that considerations having to do with mathematical simplicity incline us to use a framework which is based on the real numbers rather than one which is based on the rational numbers, even though, so far as observation and measurement is concerned, identifying points of space with respect to rational numbers would be justified. This suggests two criteria for evaluating frameworks, namely agreement with observation and mathematical simplicity. However, there are other criteria too and conflicts among criteria are possible. For example, if one is working on the solution of some problem in archaeology, let us say a problem involving classification of fossil bones, one might have to decide between two systems of classification, system one and system two. System one might be simpler to apply in most cases, but, for some cases, might not yield a decision on classification. System two might be much more difficult to apply in general but might yield decisions in a wider range of cases so that, using system two there would be fewer freaks.

I have two objections to the position Carnap has expressed in this paper. For one thing, in describing the questions with which a scientist is concerned I very often do not know whether they should be described as internal or external questions. For another, I do not see how Carnap's claim that the decision to accept a linguistic framework does not involve ontological commitment can be justified. Consider first the internal-external distinction. With respect to the archaeological example, is the archaeologist who is trying to classify the bone specimens trying to answer an internal or an external question. One can describe what he is doing as trying to decide between two linguistic frameworks—which makes his question appear external to either framework. Or one could say that he is working within a larger framework accepted by archaeologists within which his question is internal.

Let us look at this in more detail. The archaeologist is trying to classify some bone specimens. He may have to invent some species-names in order to carry out the classification. Thus he is making up a language. The decisions concerning what names to use would appear, within Carnap's way of thinking, to be answers to external questions. But, he wants to arrive at a classification which reflects a real distinction. In particular, if he decides that two bone specimens came from two different animals and he gives different species-names to each animal it reflects his belief that the animals belonged to distinct populations which were not interbreeding. The decision regarding species-names reflects a factual belief and clearly the statements made using the new species-name, statements such as "This bone is the jaw of an ..." are factual statements. Carnap's distinction between internal and external questions seems misguided. The archaeologists original question, namely, "What species-names should be used in classifying these bone specimens?" may be external to the linguistic framework at which the archaeologist finally arrives after having made his decision. But the question is not merely practical as opposed to theoretical or cognitive? The answer to the question (expressed in the language which the archaeologist finally adopts) is a factual statement, and not simply an expression of a decision to choose one linguistic framework rather than another.

That the distinction between external and internal questions with reference to linguistic frameworks is misguided can be shown also with respect to a mathematical example. Suppose that a mathematician is wondering whether some theorem which is true for real numbers also holds true for more general topological spaces. In order to answer this question, the mathematician may have to introduce some new mathematical terminology. Thus he has to adopt a linguistic framework. In particular, suppose he has to adopt a linguistic framework containing a generalized concept of continuity. The mathematician may raise the question "How should 'continuity' be defined?" This question is external to the new linguistic framework. But the answer to it is arrived at in light of cognitive considerations. If he chooses to define continuity in a particular way this reflects certain factual beliefs. In particular it reflects the belief that functions which were continuous under the original definition of "continuity" will remain continuous under the generalized definition. Again, the question "How should 'continuity' be defined?" is not practical as opposed to cognitive.

Carnap wanted to claim that the question "Does the system of numbers exist as a whole?" is a pseudo-question because it is neither external-practical nor internal-cognitive. But this claim seems to me unjustified. I have been trying to show that there are perfectly legitimate questions which are neither purely external-practical nor internal-cognitive. The question "Does the system of numbers exist as a whole?" may be such a question. To conclude that it is merely a pseudo-question is unwarranted.

The internal-external distinction may be appealing to some people due to confusion of the distinction drawn by Carnap with other distinctions. For example, one could draw a distinction between questions which are internal to scientific *disciplines* and questions which are external to such disciplines. The question "How should continuity be defined?" would appear to be internal to the discipline of mathematical analysis. Whether some question is internal or external to a discipline will clearly depend on how the notion of a discipline is defined. And the decision concerning how this notion is defined will reflect theoretical considerations.

We tend to think of different sciences as different disciplines. And, as such, they include not only different theories concerning reality but also different methodologies for validating those theories. With respect to this distinction a distinction between valid questions and pseudo-questions might be drawn. For example, if someone were to ask for a mathematical proof of the existence of some early species of man, he might be asking a pseudo-question—a question which cannot be answered because to answer it would involve an inappropriate transfer of the methodology of one discipline (mathematics) to validate a theory within another discipline. However, if we are inclined to make this distinction between pseudo-questions and real questions we must use it with extreme caution. Transfer of methodologies or theories between disciplines is not all that uncommon and often leads to fruitful results. It should not be blocked simply to preserve some sort of logical elegance in one's theory about science.

The second criticism of Carnap's position is that his conclusion that the decision to accept a linguistic framework involves no ontological commitments seems to me unjustified. If his position were correct, it would follow, for example, that a scientist who adopted a scientific framework containing the predicate "—is a star" and further which contained the claim "there are stars" would not be committed to the existence of stars even should this claim be validated within the rules of the framework. In Carnap's paper he was primarily concerned to show that the use of linguistic frameworks in semantics in which one refers to properties does not commit one to the assertion that there are properties even if the statement "There are properties" were derivable within the framework. But this conclusion seems to me to be mistaken. The astronomer is committed (ontologically) to the existence of stars and the semanticist in question is committed to the existence of properties. Carnap has not succeeded in providing a way in which one could accept mathematical principles and still remain uncommitted to the ontological implications of such principles.

NOTES

1. In presenting this semantical system we have not followed Carnap's example in detail in (10).

2. In the discussion which follows we have tried to simplify Carnap's presentation. I do not believe that the modifications introduced affect the philosophical issues.

3. Carnap, (10) p. 34.

4. The use of variables distinguished with respect to levels is normally accomplished by adding numerical subscripts or superscripts to the predicate variables.

The levels are also referred to in some works as types. The introduction of level or type distinctions was intended, of course, to avoid contradictions such as Russell's paradox.

5. Carnap, (10) p. 33.

6. For critical discussion of the verification theory of meaning see Scheffler, (85) part II.

7. A number of different axioms of infinity have been adopted by different set theorists. Sometimes it is stated in the following way. One assumes that the set whose only member is x is not identical with x. Then the axiom of infinity may be stated as follows: There exists at least one set S such that 0 is a member of S and such that if any x is a member of S then the set whose only member is x is also a member of S. See Fraenkel, (18). This axiom implies that these exists a set with infinitely many members.

8. For extensive discussion of this principle and further references see Chihara, (14).

9. Russell, (81) p. 63.

10. Poincaré, (62) p. 45 f.

11. For an explanation of the ramified theory of types and contrasts with simpler theories of types see Copi, (15) and Hatcher, (26).

12. The explanation of predicative functions which I have given is based on Russell's explanation in (81) pp. 77–8.

13. Russell, (81) p. 78.

14. In this work I am assuming that the reader is familiar with these paradoxes and so have not stated them. Russell gives these paradoxes and others in (81). They have been explained in many works.

15. Russell, (81) p. 82.

16. Ramsey, (75).

17. See Carnap, (9). Quine has also defended the use of impredicative definitions. He claims that they are not circular. See (73) p. 242.

18. Ramsey, (75).

19. For discussion of such logics see Hatcher, (26) or Quine (73).

20. Carnap, (13).

21. See Quine, "On Carnap's Views on Ontology" in (72).

22. Carnap, (13) p. 234.

23. *Ibid.*, 241.

24. *Ibid.*, 241.

25. *Ibid.*, 242.

PART II

Chapter Five

Fictionalism, Proof, Gödel's View of Mathematical Knowledge

31.   There is another philosophical position which is similar to those that we have considered above in that it denies the existence of the queer entities referred to in mathematical statements. However, this view differs from those other views in that it allows that mathematical statements really do assert or imply the existence of mathematical entities. According, to the view in question, such mathematical statements are false. The entities referred to in mathematical statements do not exist. Further, according to the view in question, mathematical statements are not merely false, they are known to be false.

One exponent of the view in question is Hans Vaihinger. According to Vaihinger, mathematicians deliberately make false assertions where doing so contributes to the development of a theory of greater generality. In developing his view he was led to distinguish between statements of fictions and statements of hypotheses in science. The basic goal of science is to formulate hypotheses. These are statements which are "directed towards reality." The intention of the users of these statements is that they "be proved true, real, and an expression of reality."[1] However, in the development of scientific theories which are sufficiently simple and general for understanding and for applicability to scientific problems it may be necessary at times to make statements which we know are not true. Such statements are, according to Vaihinger, "fictions." They are not intended to be taken as true and are not subject to processes of verification. As an example, Vaihinger mentions Goethe's idea of an "animal archetype." According to Vaihinger, Goethe did not want to assert the existence of such an archtype. He merely wanted to say that animals could be regarded *as if* they were modifications of one basic form.[2]

Vaihinger gave many examples of the occurrence of fictions in mathematics. The objects referred to in mathematical statements, he alleged, not only do not exist but are often self-contradictory.

He claimed, "Negative numbers are self-contradictory as all mathematicians admit; they are the extension of subtraction beyond the logical possibility of its application. Fractions are a product of the same method in division and so are the irrational numbers in taking roots. The most preposterous of these number constructs are the imaginary numbers, and the constructions given to them by Gauss, Drobisch, and others have in no way altered their fictional and contradictory nature."[3] This quotation suggests that Vaihinger at least regarded the natural numbers as existent. However, this is apparently not the case. Statements regarding the natural numbers were to be regarded as "semi-fictions." Statements of this sort contradict reality though they are not inherently self-contradictory. In other words, statements which imply the existence of natural numbers are simply false.

At first reading some of Vaihinger's statements may strike us as preposterous. Nonetheless we should not simply dismiss his point of view. Fictionalism embodies a certain view concerning mathematical knowledge which many people have found plausible thus it is appropriate that we begin our discussion of mathematical knowledge through a consideration of the fictionalist point of view.

Vaihinger held that fictions are ultimately eliminable from the corpus of scientific theory. Indeed he held this to be a goal of scientific theorizing. Nonetheless, he thought, at a certain stage use of ficitions is justified by expediency. Fictions are expedient in the development of simple and general theories in science. Such theories enable us to deal with a wide range of phenomena, that is to say, to remember and organize data and to bring it to bear in the explanation of specific facts. Fictions which are expedient at one time may subsequently cease to be expedient and so should be discarded at that later time. Just how Vaihinger might have anticipated the elimination of mathematical concepts from science however is not clear. Perhaps, he would have found a way to interpret the development of logical theory in the works of Frege or of Russell and Whitehead as providing a way of eliminating statements about numbers and other queer entities.

There is clearly some validity in Vaihinger's remarks. It is not unusual for mathematician's to believe that some of the concepts that they introduce are fictions. For example, Leibniz in a letter stated:

> Expressions like "Extremes meet" go a little too far, e.g., when we say that the infinite is a sphere whose center is everywhere and circumference nowhere. Such expressions must not be taken too strictly or

literally. Nevertheless, they still have a particular use in discovery, something like that of imaginaries in Algebra ...[4]

Again, in a recent introductory text in topology the author wished to introduce the system of extended real numbers. This system includes all the numbers in the system of real numbers plus two other numbers, namely a greatest lower bound and a least upper bound for the reals which are normally represented by the symbols $+ \infty$ and $- \infty$. That the author must regard these numbers as fictions is suggested by some of his remarks and by the fact that he several times refers to these two numbers as symbols. He seems here to confuse the symbols with the object symbolized.[5]

Of course, the most well-known examples of the introduction of contradictory concepts into mathematics is the introduction of infinitely small numbers (infinitesimals) in works on differential calculus. This concept entered into assumptions by Leibniz and others to the effect that there are infinitely small differences between quantities. If there is an infinitely small difference between two quantities A and B then the difference between A and B is not equal to zero. However, Leibniz, De l'Hospital and others assumed that in cases in which there is an infinitely small difference between quantities A and B then the difference between A and B is equal to zero. This notion is thus clearly self-contradictory. Yet Leibniz believed that the making of such self-contradictory assumptions was justified because, he thought, the reference to infinitesimals could ultimately be eliminated. He thought that such assumptions (the assumption that there exist infinitesimals) were false but that they were useful because they permitted shortening of otherwise tedious arguments and statements and also because they contributed to the discovery of truths.[6] This is precisely the doctrine which Vaihinger apparently wished to extend to all mathematical concepts, namely that false and even self-contradictory statements are justifiable in mathematics due to their usefulness.

Vaihinger's claim that all mathematical statements are false is, I believe, untenable. Few mathematicians today, if any, will admit that negative, irrational and complex numbers are self-contradictory. Subtraction can be defined as addition of the negative of a number and not as an autonomous operation which improperly applied leads to negative numbers. Similarly division can be defined as multiplication of the reciprocal. The finding that "fictions" do occur in mathematics as in the case of the infinitesimal or of infinite numbers does not support the contention that all mathematical concepts are self-contradictory. Contradictions have been eliminated as Vaihinger says they should. But the resulting theories continue to be theories about numbers or pairs of numbers or sets of numbers. For example, complex numbers which Vaihinger thought were "preposterous" can be explained as ordered pairs of real numbers.[7] Currently accepted mathematical theories postulating the existence of real numbers or of complex numbers are not known to be self-contradictory. Indeed, they are generally believed to be consistent.

According to Vaihinger, it is not the case that "an idea which is found to be useful in practice proves thereby that it is also true in theory."[8] In other words even though an assumption is useful in facilitating discovery or in solving problems, one may not conclude that the assumption is true. The assumption of the existence of square roots of negative numbers facilitated the solution of equations all of whose roots are real numbers. But, so long as the concept of a negative square root was self-contradictory one could not allow that the usefulness of assuming the existence of such numbers established the truth of that assumption. Self-contradictory statements simply cannot be true and so cannot be established as true. In this we can agree with Vaihinger. However, we cannot go along with Vaihinger that all mathematical statements are false or consequently that there is no mathematical knowledge. As the author of a recent work in philosophy of mathematics says "I regard as a datum the assumption that most people know some mathematical truths and some people know many."[9] Vaihinger's view is incompatible with this datum and so I must reject it.

My rejection of Vaihinger's position does not rest simply on the fact that his view is incompatible with what I take to be a datum or fact. I believe that the pragmatic theory of knowledge which he rejects is in fact valid. This is not to say that I hold that a self-contradictory statement is established as true if it is "useful." In order that a statement be true it is necessary that it not be self-contradictory. That is self-consistency is a necessary condition of truth for statements. But this leaves open the possibility of regarding consistent statements which prove useful in the elaboration of mathematical theories which are, in turn, useful with respect to other areas of knowledge, as thereby being validated or confirmed. For example, the postulation of negatives and reciprocals in the theory of real numbers is apparently consistent and, since real number theory is useful in the development of calculus as well as in applications of mathematics in natural science, we may regard the postulation of negatives and reciprocals as being confirmed.

The pragmatic theory of knowledge which we have just sketched is rather controversial. Many philosophers would not subscribe to it. In order to justify acceptance of this theory it is necessary that we critically review alternative theories of mathematical knowledge. It is to this task which we now turn. We will commence our investigation by considering the concepts of mathematical proof and mathematical intuition.

32. Mathematical Proof

The most striking thing about mathematical knowledge is that so much of it is obtained from mathematical proofs. The statements which mathematicians accept as true are, for the most part, statements which have been proved. There are, of course, historical examples in which many mathematicians have accepted as true statements which they subsequently come to realize are false. As an example Imre Lakatos in his paper "Proofs and Refutations" discusses Euler's conjecture that for all polyhedra the number of vertices plus the number of faces equals the number of edges plus two.[10] In the nineteenth century many mathematicians thought that this conjecture was true because they thought that it had been proved by Cauchy. Subsequently it was discovered that the theorem is false. It is clear, however, irrespective of such examples, that not all statements can be known as the result of a proof. This is readily seen from a consideration of the nature of a proof. It is to this task which we now turn. In light of the complexities involved in formulating a definition or explanation of the nature of mathematical proof, it is somewhat surprising that so little has been written by philosophers and mathematicians concerning this topic.

To start our explanation we shall say that a proof is an argument but not all arguments are proofs. By an argument we mean a set of statements one of which is the conclusion and the others of which are the premises. The premises are the reasons which allegedly support the conclusion. We can identify the premises as the statements which are offered as supporting or proving the conclusion. Clearly, we cannot simply define a proof as an argument in which the premises support the conclusion. There are two reasons for this. For one thing the expression "support the conclusion" is metaphorical and, as such, stands in need of explanation itself. Secondly, if "support the conclusion" is understood as "establish that the conclusion is true" then to say that a proof is an argument which supports its conclusion is to give a circular definition of *proof*. Saying that a set of statements establish a conclusion as true is just another way of saying that the statements prove the conclusion.

These considerations suggest however that in distinguishing proofs from other arguments we should consider the structure of the argument itself, its relationship to reality and certain relationships of the argument to human beings. We shall call the latter relationships in question *epistemic* relationships. Let us consider first the structure of the argument itself.

With respect to the structure of the argument itself, it is customary to consider as a class those arguments which guarantee that if their premisses are true then their conclusions are true also. Such arguments are said to be *valid*. Valid arguments are arguments which cannot possibly have both true premisses and a false conclusion. Arguments of this class, that is valid arguments, are distinguished from arguments which do not guarantee that if their premisses are true then their conclusions are true but which are nonetheless reliable in that if their premisses are true then in most cases their conclusions will be true also. Let us call arguments of this second category *reliable*. The traditional problem of induction may be understood as the problem of trying to establish that arguments of certain forms, namely inductive arguments, are reliable. Following up a suggestion of Nicholas Rescher in his book *The Coherence Theory of Truth* we may distinguish a third category of arguments.[11] Arguments of this class are not classified with respect to their capability to transmit truth from premisses to conclusion. Rather, such arguments are thought of as ferreting out truth from among a set of statements which are to be considered as "truth-candidates." That is to say, with respect to arguments of this category, the premisses are not truths which can establish the truth of the conclusion. Rather the premisses are statements which may be true or which may legitimately be considered as truth-claims. An argument in this sense is an analysis of these truth-claims which supposedly reveals any truths which may be contained in the set of truth-claims. Rescher, of course, argues that the appropriate type of analysis involves considerations having to do with the coherence of subsets of the set of truth-claims.

Needless to say, this classification of arguments is not

exhaustive. Many other forms of arguments could be distinguished, e.g., arguments which guarantee that if their premisses are true their conclusions are false, arguments which are unreliable, arguments which are persuasive but unreliable, etc. However, we shall not discuss arguments of any of these other sorts. It is clear that mathematical proofs are supposed to be arguments of the first sort discussed above, namely they are supposed to be valid arguments.

As noted earlier, valid arguments are arguments which cannot possibly have true premisses and a false conclusion. In recent years some philosophers have thought that we (human beings) were in possession of infallible means for distinguishing valid arguments from all other types of arguments. Valid arguments, it was alleged, could be distinguished by reference to their "logical form" alone. I do not believe that this claim can be defended against criticism. The philosophers who defended this claim seem to have thought that the method of logical analysis in which one represents arguments as instances of formula of such first-order quantificational logics as those found in many modern textbooks of logic was "the correct method of analysis" and that the formulas which the argument instantiated represented *the form* of the argument. But development of alternative logics and alternative set theories has shown that this claim is untenable. These alternative logics provide alternative ways of representing the structure of arguments. Furthermore, using one of these logics as a basis for representing the logical form of an argument often involves translating the statements of the argument into other statements which instantiate the appropriate forms. In many cases different translations of the same statement are possible even within the framework of one system of logic. Thus, a statement and so also an argument may have many different forms. The number of possible forms is, of course, increased by consideration of forms within different logical theories. In light of these considerations it seems a mistake to speak of *the form* of an argument.

Further, the choices one makes with respect to translation and logical theory affect the decision as to whether the argument is valid. A given argument, translated in one way with respect to one logical theory may be valid whereas the same argument translated differently or translated with respect to a different logical theory will be invalid. For example, if the following argument is translated with respect to first-order quantificational logic without identity then it is invalid. However, if the logic includes identity then the translation will be affected and the argument proves valid. Intuitively, the argument appears valid and so the latter translation seems more correct. The argument is

> John is tall.
> John is Peter.
> Thus, Peter is tall.

Or again, the following is an example of an argument which intuitively is invalid but when translated into the standard first-order logic of quantification theory (indeed into the propositional part of such logic) it acquires a valid form:[12]

> It is not the case that if you drink milk you will get intoxicated. Therefore, you will drink milk and will not get intoxicated.

Or again, the following is an example of an argument which intuitively is valid but when translated into the standard first-order logic acquires an invalid form:[13]

> Tom, Dick and Harry are shipmates. Therefore, Tom and Harry are shipmates.

These considerations suggest that making a correct determination regarding the validity of an argument requires some sensitivity to meaning as well as skill in using the range of logical methods of analysis available. Determining the validity of an argument is not, therefore, a purely mechanical or formal matter. It is a matter with respect to which intelligent and rational men may make mistakes.

Our discussion of the relation of an argument to reality can be somewhat briefer than our discussion of logical structure. In order for an argument to be a proof, its premises and consequently its conclusion must be true, that is what such statements affirm or deny must be as they say. This condition is imposed by the fact that a proof is supposed to yield knowledge. As a result of a proof we come to know the conclusion of the proof. If the conclusion were not true then we could not know it.

Are the above two criteria sufficient for distinguishing proofs from other arguments: That is, could we define a proof as a valid argument with true premises? I think not. I believe that in order to correctly distinguish mathematical proofs from other arguments we must consider what we have called "epistemic relationships", that is, relationship of the argument to the potential knower.

That the definition of *proof* as a valid argument with true premisses is unsatisfactory can be readily seen from the fact that this definition does not distinguish circular arguments from proofs. But the only reason for distinguishing circular arguments from proofs is that a proof is intended as a means by which we may come to know something new. The new knowledge is the conclusion. The premisses represent old knowledge. If the conclusion is also a premise, or if in order to know some of the premisses one must first know the conclusion, then the alleged proof fails with respect to its intended purpose, that is, it fails to be a means by reference to which we can acquire new knowledge.

Further epistemic relationships must also be considered. It is not sufficient to consider an argument a proof if it is merely non-circular, valid and has true premisses. The logical relationships between the premisses and the conclusion in virtue of which the argument is valid must be discernible by human minds. That this is so is shown by the following considerations: Suppose that T is some statement of real number theory and suppose further that T is a valid logical consequence of the axioms of real number theory but that no one has ever been able to deduce T from those axioms. Then we would not consider the following argument a proof even though it may indeed have true premisses, is valid and non-circular:

    1. Axioms of real number theory
    2. If axioms of real number theory then T
Thus,
    3. T.

In claiming that the logical relationships between premisses and conclusion of a proof must be discernible by human minds I am only reiterating a point made by Locke concerning the nature of demonstrative knowledge. According to Locke, in such knowledge

> "when the mind cannot so bring its ideas together as by their immediate comparison ... to perceive their agreement or disagreement, it is fain, *by the intervention of other* ideas ... to discover the agreement or disagreement which it searches."[14]

Locke recognized that there could be cases in which it would be impossible that an argument's premisses be true and its conclusion false but that due to the great complexity of the ideas involved human beings could not recognize that the argument was indeed valid. In order for an argument to be a proof such must not be the case.

One final epistemic relationship must be noted. Valid logical relationships are not such as to generate knowledge from premises which are not known to be true. That is, the conclusion of a valid deductive argument is not known to be true unless the premises of the argument are known to be true. And so an argument is not a proof unless its premises are known.

The above definition of mathematical proof makes proof relative to what people under certain circumstances come to know. It may be asked whether such a relational analysis of the nature of proof implies that mathematical truth is also relative to human cognitive ability. Our analysis would indeed have this consequence if we also held that a mathematical proposition is not true unless it has been proved. We could then argue that a proposition A is true only if A is proved and that A is proved only if there is a valid argument whose premises are known to be true, and from this we could conclude that A is true only if there is an argument whose premises are known to be true. However, since I do not believe that mathematical truth is relative to what is known, I do not accept the proposition that a mathematical proposition is true only if there is a proof of it. Pythagoras theorem would have been true even if human beings had never evolved and so had never come to know it.

It must also be noted that our definition of proof presupposes that we may understand the nature of mathematical knowledge independently of our understanding of the nature of proof. If this is not the case then our definition of proof is still circular. However, I do not believe that our definition of proof is circular in this way. In support of this I am not prepared to offer a definition of *mathematical knowledge*. But I think that the following argument has some weight. The most plausible suggestion which implies that the concept of knowledge is dependent on the concept of proof is the suggestion that if a person knows a proposition P then he knows a proof that P is the case. Now, if to know that P entailed knowing a proof that P, then it would be inconsistent to claim that there are mathematical propositions which are known without proof. But this does not strike me as inconsistent. Further, most philosphers who have developed theories of mathematical knowledge would not have found this inconsistent either since all major theories concerning mathematical knowledge allow that there are some propositions which are known without having been proven. I conclude that, in all probability, it is not inconsistent to hold that there are mathematical propositions which are known without proof and that the concept of knowledge is independent of the concept of proof.

Some philosophers have argued that knowledge should be defined as follows: *A knows that P* means that *P* is true, that *A* believes that *P* and that *A* has completely adequate evidence that *P*. If this definition were applied to mathematics, it would be tempting (to some people perhaps) to understand the expression *completely adequate evidence* as meaning *proof*. In light of the above considerations we would have to reject this way of construing knowledge at least in mathematics. While a proof that *P* may be completely adequate evidence that *P*, it is not the case that completely adequate evidence that *P* is, in all cases, a proof that *P*.

It should be noted that while the above explanation of the nature of proof in mathematics is intended to reflect what mathematicians mean when they speak of or think of a proof, it is not the case that all arguments which are referred to as proofs in mathematical works would satisfy all of the criteria we have mentioned. Arguments as stated in texts and referred to as proofs may fail to be valid arguments. In some of these cases of course this failure is especially significant. These are cases in which the mathematicians believe that the argument is indeed valid when it is not. Arguments which are intended as proofs and which are believed valid but which are not valid are called fallacies. Surely it is a virtue and not a defect of our analysis of proof that it does not classify fallacies as proofs. In other cases, I believe, that the mathematicians are aware that the arguments as stated are not valid. In these cases for purposes having to do with economy of exposition or clarity of thought the mathematicians have deliberately omitted information which is judged to be commonly known. Thus the stated arguments should be considered as enthymemes. In cases in which the argument in a mathematical work is called a proof but is really an enthymeme we can perhaps understand the stated proof as an abbreviated version of the proof itself. We might say that the stated proof is not intended to be taken as the proof but rather as a set of statements which enable the mathematician to develop the whole proof for himself if he is so inclined.

Given that some mathematical knowledge is obtained via proofs, there must be some mathematical knowledge which is otherwise obtained. The basic problem in regard to theories of mathematical knowledge is to give a satisfactory account as to how such other knowledge is obtained. In succeeding sections we shall consider various theories concerning this problem. We shall begin these considerations in the next section.

33. The central problem regarding our knowledge of mathematics arises in the following way: If a mathematical theorem is known to be true as a result of a proof, that means that the theorem is the conclusion of an argument. With respect to the types of arguments in mathematics, if a theorem is known because it is the conclusion of an argument then the premisses of the argument must be known to be true also. Of course, the premisses might themselves be conclusions of still other arguments. But, we can see that we are faced with the following alternative. Either there are some mathematical premisses which are known to be true without proof or the process of proving statements true on the basis of additional arguments is interminable. In this latter case we would say that the number of premisses of a mathematical proof is infinite.

The second of these alternatives, that is that there are proofs with an infinite number of premisses has seemed to most people to be mistaken. This view might be supported by saying that an argument with an infinite number of premisses cannot be a proof. Someone who thought in this way might argue that if an argument is to be a proof then someone must be able to come to know the conclusion as a result of knowing the premisses and understanding them and that it is impossible for a person to know and understand an infinite number of premisses. But this argument does not strike me as being sound. It just is not clear that it is impossible for a person (with a finite mind) to know an infinite number of premisses. It may be impossible for a person to say an infinite number of things in a finite time or for a person to actually think one by one of each of an infinite number of premisses in a finite amount of time. But it is not clear that having knowledge of an infinite number of premisses entails being able to say them or to consider them one at a time. It may simply be that there is an infinitely large set of statements of which the person knows each one. This does not seem impossible at all. A person easily knows an infinite number of trivial consequences of anything that ke knows. For example, if he knows a statement 'S' then he knows the statement "S or S'", and the statement "S or S' or S''", etc.

If we could argue that people have mathematical knowledge and that knowing an infinite number of premisses was impossible then we could certainly conclude that the second alternative mentioned above was untenable and so would have to accept that there are some mathematical premises which are known to be true without proof. However, even though we cannot accept the claim that knowledge of an infinite number of premises is impossible, we still reject the second alternative since, in fact, all mathematical proofs are finite. When someone knows a proof of a theorem he knows a finite number of statements. He can give all of the statements of the proof in a finite time or space. Thus, we conclude, that if there is mathematical knowledge then some mathematical principles are known without proof.

Several positions come to mind at this point. One is that there is indeed no mathematical knowledge. This position seems to me to be untenable. Subsequently I shall try to indicate what premises might lead a person to such a view and shall also indicate why I think those premises are mistaken. The second view is that knowledge of mathematical premises is non-empirical. Variations of this sort of view are platonism, intuitionism and some versions of formalism. We shall consider such views in succeeding sections. The third sort of view is that knowledge of mathematical premises is empirical. We shall develop this view in a later part of this book. It is, I believe, essentially correct. We shall begin our study of theories of mathematical knowledge in the next section with a consideration of Platonism.

34.   There are some facts which render the view that some mathematical knowledge is (a) unproved and (b) acquired in a non-empirical way plausible. Consider a well-known example. A teacher is alleged to have given the mathematician Gauss the following mathematics problem when he was a student in an elementary class. The teacher wanted to keep the class busy. The problem is to find the sum of all the numbers from one to one hundred. Gauss allegedly solved the problem very quickly by seeing that the sum in question equals one hundred and one multiplied by fifty.[15] It might be alleged that what led Gauss to a correct solution to this problem was an awareness of the deep structure of the sum involved and that clearly such awareness is not a form of sensory observation. No doubt many teachers of mathematics have observed similar phenomena in their students. Such examples of apparently non-observational learning suggest that some people at least have the ability to grasp mathematical truths directly, that is, in a non-sensory way. Such non-sensory awareness of a mathematical structure may be called mathematical intuition.

The existence of such intuition is also suggested by examples of the discovery of mathematical truths by advanced researchers. Such workers are often studying non-observable objects such as functions, sequences, or topological spaces. Then seem to develop a kind of familiarity with the properties of the objects in question. When they discover that the objects have some property it often seems as if they perceive the objects. Since the objects are non-observable it is not implausible to speak of non-sensory awareness of the properties of such objects. Consider for example a mathematician who is trying to solve a problem which involves finding the sum of an infinite series. After reflecting upon the series for a period of time he becomes aware that the terms of the series fall naturally into certain groups and that in light of this the sum is readily determined. The awareness that the terms of the sequence fall into certain groups and that this leads to a solution to the problem appears to be a non-sensory awareness. But, if such awareness is possible in problems such as these then why cannot it be relied upon as a basis for knowledge claims? Certain philosophers have indeed maintained that such awareness is the basis for our knowledge of the axioms of set theory. One such person is Kurt Gödel.

Gödel maintained that it was quite meaningful to raise questions concerning the truth or falsehood of the axioms of a consistent mathematical theory and that this question with respect to certain set theories might be settled by "a perception ... of the objects of the theory."[16] He said

> But, despite their remoteness from sense experience, we do have something like a perception of the objects of set theory, as is seen from the fact that the axioms force themselves upon us as being true. I don't see any reason why we should have less confidence in this kind of perception, i.e., in mathematical intuition, than in sense perception, which induces us to build up physical theories and to expect that future sense perceptions will agree with them.[17]

Gödel's argument may be developed along the following lines. Suppose that I am looking at the sky on some sunny day when there are no clouds. Then, I am having some sensory experiences which are such that the statement "The sky is blue." forces itself on me as being true. In an analogous way, when I am considering sets I may have experiences which are such that certain axioms of set theory force themselves upon me as being true. For example, when I think about sets I may have experiences which force the following statement upon me as true: For any set S there is a set T

such that an object t is a member of T if and only if t is a subset of S.

One obvious sort of objection to make against Gödel's suggestion is that intuition is not sufficiently reliable to be a source of knowledge. Intuitions are frequently mistaken. The forcefulness of this objection, with respect to the axioms of set theory, can best be felt by reflecting on the axioms assumed by Frege, namely the axiom of extensionality and the axiom of abstraction. The axiom of extensionality states that for any sets x and y if x is identical to y then P is a property of x if and only if P is a property of y for any property P. That is, the axiom of extensionality affirms

$$(\forall x) \, (\forall y) \, (\forall P) \, ((x = y) \supset (Px \equiv Py))$$

The axiom of abstraction says that there is a set Y consisting of all objects x of which P is true for any property P. That is, it affirms that

$$(\bar{\exists} y) \, (\forall P) \, (\forall x) \, (x \in y \equiv Px)$$

These axioms strike most people as highly intuitive. Surely, it appears that for any property P there is the set of all and only those things possessing P. Some people might object to this axiom (abstraction) by affirming that it may be that nothing possesses P. But, in set theory as normally understood, to say, for some property P that nothing possesses P, does not imply that the set of all and only things possessing P does not exist. If nothing possesses P then the set in question is the empty set. Yet it is well known that axiom of abstraction is inconsistent. From this axiom we can derive Russell's paradox, namely x is a member of y if and only if x is not a member of y that is,

$$x \in y \equiv x \in y$$

The defender of Gödel's view can reply to this objection in the following way. The objection is that intuition is not a source of knowledge because intuitions are sometimes mistaken. But, you might as well argue that sense perceptions are not a source of empirical knowledge since sense perceptions sometimes are mistaken. That is, just as a person on the basis of his sensory experiences may feel forced to accept certain statements which subsequently prove to be false, so also a mathematician may feel forced to accept certain principles which subsequently prove to be false. But this fact does not show that intuition never yields knowledge just as the falsehood of some empirical statements doesn't show that sense perception never yields knowledge. Indeed Gödel has made essentially this claim. He said, "The set-theoretical paradoxes are hardly any more troublesome for mathematics than deceptions of the senses are for physics."[18] Pushing the analogy with sense perception a bit further we might say that just as statements based on sensory evidence are not known with certainty to be true so also statements based on mathematical intuition are not absolutely certain. Nonetheless, the defender of Gödel's view may hold, mathematical intuition has some epistemic value. That is, the existence of mathematical intuition is strong evidence that the statement is true. He might even say that a statement affirmed on the basis of mathematical intuition ought to be considered as true unless it is known to be self-contradictory. (Just as in physical sciences one might hold that statements based on sensory evidence ought to be accepted as true unless one has good reason for supposing them false.)

A second objection to the view that intuition provides evidence for the truth of mathematical statement is, it might be said, that to let mathematicians decide which axioms are true on the basis of their intuitions is nothing more than letting them decide what is true on the basis of their taste. To use this method of discovering the truth is, in effect, to use what C.S. Peirce called "the a priori method of" fixing beliefs.[19] But then we cannot expect, so this objection runs, that the beliefs arrived at in this way will correspond to any objective realities. That is to say, we have no reason to regard a belief as true unless it is "determined by nothing human, but by some external permanency—by something upon which our thinking has no effect."[20] We may be caused to believe that the axioms of a set theory are true by some social circumstances. That they are regarded as true may be a result of some myth to which we have been socially conditioned. Thus we should not let the fact that we feel forced to accept these axioms count as evidence for their truth.[21]

The defender of Gödel's view may respond to this objection in the following way. He may suggest that it is quite incorrect to say that the appeal to intuitions in mathematics is nothing more than settling on beliefs on the basis of inclination or taste. He might claim that intuitions regarding the truth of some axioms can be confirmed by reference to other mathematical truths. Indeed Gödel has taken this position. He said

... even disregarding the intrinsic necessity of some new axiom, and even in case it has no intrinsic necessity at all, a probable decision about its truth is possible also in another way, namely, inductively by studying its "success". Success here means fruitfulness in consequences, in particular in "verifiable" consequences, i.e., consequences demonstrable without the new axiom, whose proofs with the help of the new axiom, however, are considerably simpler and easier to discover, and make it possible to contract into one proof many different proofs. The axioms for the system of real numbers, rejected by the intuitionists, have in this sense been verified to some extent ...[22]

Gödel is here suggesting that axioms in set theory might receive a kind of verification similar to that accorded to principles in the natural sciences. Such verification has the following form: Starting with a general theory one deduces that under such and such circumstances certain observations may be made. Then one contrives a situation to fit the "such and such" circumstances. If, in that situation, the observations are indeed made, that provides some confirmation for the theory. The theory is not proved conclusively on the basis of those observations. It is however accorded some non-conclusive (inductive) confirmation. That is, a theory is inductively supported if it entails some statements which are independently supported. The theory in question is also supported if it renders into one coherent system a number of principles that were otherwise unrelated to each other.

Now, Gödel is suggesting that in mathematics axioms of set theory may be confirmed in an analogous way. He suggests that the axioms of real number theory may be confirmed if, on the basis of these axioms, one may deduce principles which are independently known to be true. Suppose for example, that there were many principles of calculus which had been proved in a range of different ways. The claim that there exists a complete ordered field, i.e., the claim that axioms of real number theory are true, might be confirmed if on the basis of this claim we could deduce these other principles. This claim would be further supported if the proofs of these other principles were simplified or if a number of proofs could be contracted "into one proof".

Of course, one may ask how the principles known independently of the axioms of real number theory (or of set theory) are known. Presumably the defender of Gödel's position would allege that the premisses through which such principles were proved were also known by intuition. Thus, Gödel's argument may not be very persuasive to one who is prone to doubt all intuitions. What Gödel is claiming in this last paragraph is that principles which are independently supported by intuition can by mutually supporting.

In this regard we can compare his position to that of someone who is defending the validity of empirical knowledge against a critic who challenges sense perception as a basis for knowledge. In this case also no conclusive proof can be given to refute the skeptic. The best one can do perhaps is to point out that independent sensory confirmation can be given for principles which are logically related and that in this way some sense perceptions, as it were, support others. Gödel's view vis à vis the skeptic (in regard to intuition) seems no weaker in this regard than that the defender of sense perception would be in regard to arguments advanced by one skeptical about the validity of empirical knowledge.

There is still the objection, suggested by Peirce's remarks, of whether we have reason to believe that the intuitions signify any external reality. Of course, if the critic of Gödel insists that in order to have reason to believe that the axioms are true, our acceptance of them must be determined by "nothing human" then we can reject his standards as too high. Even in the natural sciences it apparently is the case that "human factors" enter into the determination of our beliefs in some respects. This is not the place to discuss such determination at length. But surely our beliefs in physics or biology are often conditioned by accidental cultural factors in various ways. If this fact does not justify total skepticism with respect to theories in the natural sciences then it will not do so either in regard to mathematics. Still, there is an important objection underlying Peirce's remark here. It is this objection which is most damaging with regard to Gödel's view. The objection, as I understand it, is this: There is no reason to think that the axioms of real number theory or of set theory, which our intuitions lead us to accept as true, correspond to properties or relations among numbers or sets, because there is no reason to think that numbers or sets are causes (or partial causes) of our intuitions. On the other hand we have good reason to think that the statements of natural science which we make concerning physical objects correspond to properties of or relations among physical objects since such objects are causes (or partial causes) of our sensory perceptions. For example, consider observation reports of capillaries as seen under a microscope. We have good reason to think that there are capillaries since capillaries clearly are part of the cause of the sense-perceptions of the observer.

The Gödelian is in no position to make the same sort of claim about intuitions that the critic makes about sense perceptions. In particular, the Gödelian cannot come up with an example analogous to the above example regarding capillaries. Suppose for example that a mathematician has an intuition which leads him to believe that Rolle's Theorem is true. (Rolle's Theorem states that for any function f defined on the real numbers which is continuous at any point x within an interval, if the first derivative of f exists at each point of the interval and if the value of f at the endpoints of the interval is zero, then there is some point in the interval such that the value of the derivative at that point is zero.) The Gödelian cannot claim that the real numbers are causing him to have the intuition which supports this statement. At any rate such a claim is clearly unjustified. Given my understanding of causal relations, there is no way that mthematical objects could be causally related to individual human minds. We shall discuss this point at great length subsequently (in connection with our consideration of the views of Mark Steiner).

Gödel, of course, anticipated even this last objection to his view. He replied to it as follows:

> It by no means follows, however, that the data of this second kind (intuitions), because they cannot be associated with actions of certain things upon our sense organs, are something purely subjective. ... Rather, they too, may represent an aspect of objective reality, but, as opposed to the sensations, their presence in us may be due to another kind of relationship between ourselves and reality.[23]

In other words, Gödel is suggesting that we must not assume that the only ways in which we can have knowledge of "objective reality" all involve actions of external objects on our sense organs. There may be some other way in which our mind could be appraised of external realities—some way not involving sensation. But this argument of Gödel's does not meet the above criticism. The critic can admit that there may be some other way in which the mind could be affected other than by means of the action of external objects upon our sense organs. He can even admit that if some sort of causal explanation relating numbers or sets to our minds (or even to our bodies) could be developed then this criticism of Gödel's position would be undermined. But, to come up with such an explanation large conceptual difficulties have to be overcome. In the critic's view, it is unlikely that the Gödelian will be able to overcome those difficulties in an acceptable manner. Failure to overcome those difficulties means that at a crucial point, the analogy between sense perception and mathematical intuition breaks down. Thus an important part of the theory of knowledge advocated by Gödel is missing and so his theory remains unacceptable.

Before completing our discussion of Gödel's view we must consider the work of Mark Steiner who, has defended the view that mathematical knowledge is obtained via intuitions from criticisms such as we have just been considering. Steiner would claim that our criticism rests on acceptance of the causal theory of perception. Steiner has attempted to show that this theory (the causal theory of perception—CTP) is unfounded.

35. Steiner has formulated two versions of the causal theory of perception. He tries to support Gödel's position by showing that Gödel's view is compatible with one version of CTP and by showing that the other version of CTP is unfounded. The first version of the causal theory of perception is formulated by Steiner as follows:

> (A) One cannot see that S ('S' replacing an arbitrary sentence) unless N ('N' replacing a name of the arbitrary sentence) must be part of a causal explanation of the perceptual experience that is part of *seeing that S*.[24]

Here the expression "seeing that" is to be understood broadly enough to include intuiting that S where S expresses some mathematical truth. According to Steiner,

> If (A) were acceptable, the Platonist could adhere to it, thus accepting a version of the CTP which is entirely compatible with the existence of mathematical "intuition." For if N be an arithmetic truth, certainly it will appear in any explanation of any perception.[25]

Steiner's argument is then that, if to accept CTP means to accept (A) then CTP constitutes no objection to the Gödelian view. The Gödelian can accept (A) since arithmetic truths will enter into "any explanation" of the occurrence of the intuition in question.

We have criticized Gödel above by claiming that for his view to be acceptable there will have to be a causal explanation in which the occurrence of intuitions is causally explained by reference to the existence of mathematical objects and that we don't believe that any such explanation can be given due to conceptual difficulties involved in causally relating mathematical objects to intuitions. This objection is not answered by claiming that any explanation of intuitions would involve arithmetical truths. Steiner is saying that if there is any acceptable causal explanation of the intuitions it will include arithmetical truths. We agree with this. Given that there are experiences which we would describe as being experiences of seeing that S, where S is a mathematical statement, it is very likely that any satisfactory causal explanation of the occurrence of such experiences will depend heavily on mathematical truths. For example, if there were a behaviorist learning theory or a learning theory based on the work of Piaget through which such experiences could be explained then it is very likely that in the explanation mathematical principles would play an essential role. However, our objection to Gödel rests on acceptance of a version of the CTP other than that expressed in (A). Steiner has expressed this second version of CTP as follows:

> (C) One cannot see an F, unless the F participates in an event that causes one to have some perceptual experience.[26]

Since we are concerned with intuition in mathematics we must understand (C) as implying that if a person becomes aware of some mathematical objects, e.g., sets or numbers, via a mathematical intuition, then there is some event in which those objects participate and that event is a cause (or partial cause) of some perceptual experience in the person. The differences between (C) and (A) are significant. This may be seen by considering an example. Consider the statement

> (S) The choice set exists.

According to (A) a person could see that (S) even though he did not see the choice set (or intuit the choice set), since he could see that (S) so long as he has some experience in which he becomes aware that (S) is true and so long as a causal explanation of the occurrence of that experience includes "(S)" in an essential way. According to (A) a person could see that (intuit that) the choice set exists even though the choice set is not a cause or partial cause of his seeing that the choice set exists. But, according to (C) a person cannot see (intuit) the choice set unless the choice set is a cause or partial cause of his seeing (intuiting) the choice set.

Understanding (C) properly requires understanding the expression "F participates in an event" in the proper way. If this expression is misinterpreted then it is possible to construe (C) as compatible with the Gödelian position. In particular if the expression "*F* participates in an event" is understood to mean that *F* is a property of the event *E*, then (C) is not incompatible with Gödel's view. For example, suppose an event *E* is an occurrence of a loud noise. Then one might say that the property loudness participates in this event *E*. Here "participation" is understood as signifying a relation that holds between abstract objects such as properties or relations and particular events. Mathematical structures or objects might indeed participate in events (in this sense of "participate"). For example, a particular event might consist in a group of twelve people being in some room at some time. The number twelve might be said to participate in this event. If "participate" is interpreted in this way then a person who saw the twelve people in the room could be said to intuit (or see) the number twelve. Clearly, in this example, intuition and sense perception are not alternative means of acquiring knowledge of objects. Intuition of abstract objects is simply sense perception of some event in which the abstract objects participate. I don't think that Gödel had this in mind when he thought that mathematical knowledge might be acquired through mathematical intuition. He apparently did not think that knowledge of mathematical truths was obtained through the perception of physical objects or events by the senses.

A second way to understand the expression "*F* participates in an event" is as follows: *F* is understood as restricted to particular substances. Universals, such as mathematical objects, are not included in the range the variable '*F*'. An event is to be understood as a particular substance or several particular substances having some property during some limited time (or standing in some relation during some limited period of time). *F* participates in an event if and only if *F* is one of the particular substances out of which the event is constituted. As an example, consider as an event a tree falling down. The tree participates in this event but the relation "falling down" does not participate (in this sense) in this event. If one understands "*F* participates in an event" in this way then principle (C) is clearly incompatible with Gödel's view since mathematical objects are not particulars (and so are not particular substances). Mathematical objects are universals (as noted in section 5 above).

But, we may ask, is it possible to reconcile (C) with Gödel's view by extending the range of *F* so as to include mathematical objects without reverting to the first way of understanding "*F* participates in an event?" This means that we must construe the event as containing some parts which are non-physical entities. For example consider the event of twelve people being in a room. The parts of this event would have to include the number twelve in addition to each of the people and the room. Suppose we construe "F participates in an event" in this third way, is Gödel's view comptible with (C)? I would say no since even if we allow the event to have mathematical parts, I do not see how we can meaningfully speak of the mathematical parts as causing (or being partial causes of) a perceptual experience. Since mathematical objects are non-physical there can be no physical process connecting a mathematical object and a human perceptual experience. Thus, if there were a causal relation between a mathematical object and a human perceptual experience it would have to be a causal relation between objects which are not connected by any physical process. (In saying that no physical process connects a mathematical object and a perceptual experience I mean for example that no electromagnetic energy is transmitted from one to the other, they are not related by another other physical forces such as gravity, etc.) The postulation of non-physical causal relations is concepatually incomptible with the methodological principles of modern science. In order to make Gödel's position compatible with principle (C) we would have to drastically overhaul our ideas regarding causal relations. This does not seem warranted merely to render Gödel's theory of knowledge compatible with principle (C). To justify such a drastic overhaul it would have to be even more implausible to regard (C) as false or more implausible to regard Gödel's theory of knowledge as false than to suppose that some causally related objects are not physically related. However, I do not believe that both of these alternatives are more implausible than supposing that there are non-physical causal relations. (C) strikes me as a reasonable principle. Therefore I am inclined to regard Gödel's view as erroneous.

Steiner attempted to defend Gödel's veiw by raising doubts concerning the validity of (C). However, I regard his suggestion that (C) might be mistaken (or inapplicable to mathematical objects) as unsatisfactory. To see why consider what is involved in the rejection of (C). To reject (C) means that there can be a person who at one time is not aware of some mathematical objects, i.e., is not intuiting them and does not know that they exist, and who, at a subsequent time, is aware of the same mathematical objects and yet this change with respect to his awareness is not brought about through any causal interaction between the objects and the person's mind. But, to suppose that such a person could exist is to suppose either that the change in the persons's awareness is uncaused or that the change is brought about through causal interaction with some other (non-mathematical) objects. But, of these alternatives, the first one is not justifiable and if we accept the second we do not have mathematical intuition but awareness of some non-mathematical objects instead. On the one hand to say that there are uncaused mental events is to assert a hypothesis that is susceptible of a good deal of empirical scientific investigation. It would be rash to accept it simply because by so doing one can preserve the theory of mathematical knowledge that we have been considering. On the other hand, suppose that the change in awareness with respect to mathematical objects is brought about by a causal interaction with other (non-mathematical) objects then there is no reason for describing the awareness in question as an awareness of mathematical objects. On this alternative, one would simply have become aware of whatever these other objects are that gave rise to the change in awareness.

36. Conclusion: In this chapter we have investigated the view that knowledge of mathematical objects is obtained by some form of non-sensory awareness of such objects. We have rejected this view primarily because we do believe that such non-sensory awareness is not possible unless mathematical objects could be causes or partial causes of occurrences of such non-sensory awareness. But we do not believe that mathematical objects could be causes or partial causes of such occurrences.

It should be noted that our acceptance of the causal theory of perception (including the causal theory of intuition) does not commit us to a causal theory of knowledge. That is we are not committed to the view that in order to have knowledge of the existence or nature of an object such an object must be a cause or partial cause of our having that knowledge. We believe that it is possible for us to have knowledge of the existence and nature of mathematical objects even though such objects are not causes or partial causes of any of our mental states. Our knowledge of such objects does not rest on a causal inference, i.e., it does not rest on an inference in which the cause of a mental state is inferred from some data.

The experience of Gauss and others does not, of course, prove that there is mathematical intuition in Gödel's sense. Gauss made an inference by which he solved the problem concerning the sum of the first one hundred natural numbers. He may have known the premisses used in this inference as a consequence of sensory experience.

NOTES

1. Vaihinger, (89) p. 85.
2. *Ibid.*, p. 86.
3. *Ibid.*, p. 57.
4. Leibniz, (45) p. 72.
5. Simmons, (86) pp. 56–7.
6. This discussion of Leibniz is based on the discussion in Abraham Robinson's book (79). There are several quotations in that book from Leibniz letters to bear out the claims we have made regarding Leibniz views, pp. 261 f.
7. For an explanation of the introduction of complex numbers which is readily understandable see, Waismann, (90).
8. Vaihinger, (89) p. 111.
9. Steiner, (87) p. 13.
10. Lakatos, (40) p. 7 ff.
11. Rescher, (76).
12. This example is based on the paper of Nelson, (58).
13. This example is due to Massey, (52).
14. Locke, (48) volume II, p. 178.
15. This example is given in many places. One such place is Brown, (7) p. 192.
16. Gödel, (22) p. 271.
17. *Ibid.*, p. 271.
18. *Ibid.*, p. 271.
19. Peirce, (59) p. 17.
20. *Ibid.*, p. 18.
21. This has been suggested by Chihara, (14) p. 67 F.
22. Gödel, (22) p. 265.
23. *Ibid.*, p. 272.
24. Steiner, (87) p. 117.
25. *Ibid.*, p. 117.
26. *Ibid.*, p. 118.

Chapter Six

Intuitionism

37. We have been concerned with the question of how an individual can justify his claim to know the premisses of mathematical proofs. The first view that we have considered is the view that such knowledge is obtained by mathematical intuition of independently existing mathematical entities such as sets or numbers. We have referred to this view as Gödel's view. It has also been called "epistemological platonism." It is one of the views in accordance with which mathematical knowledge is non-empirical. As we have seen, the major difficulty with this view concerns its failure to provide a reasonable account of any causal relationship which would explain the occurrence of intuitions in individual minds. It is natural then to consider next an epistemological position which can avoid this problem. The position that we shall next consider is commonly called "intuitionism" (though Gödel's view also could be called intuitionism). Intuitionism avoids the problem that proved too difficult for Gödel's theory by supposing that the objects of mathematical knowledge are not external to the individual mind. Numbers, etc., exist in the mind of any individal who has taken the trouble to construct them in his mind.

The most well-known exponents of this theory are L.E.J. Brouwer and Arend Heyting. Brouwer traces his theory to Kant who held that built into the very nature of the human mind are certain forms of perception. Kant thought that there were two such forms—the form of spatial perception and the form of temporal perception. Mathematical knowledge consists, Kant thought, in knowledge of these forms. Geometry is knowledge of spatial perception and arithmetic is knowledge of temporal perception. Brouwer eschews part of Kant's thought on this matter. He considers that the discovery of non-Euclidean geometries completely invalidates Kant's views concerning our knowledge of geometry.

According to Brouwer "the fundamental phenomenon of the human intellect" consists in human experience being separated into distinct parts or units which are nonetheless united in consciousness.[1] We are aware of a sequence of distinct units of experience which occur to us successively. We are also aware that these distinct units are elements of one consciousness. Brouwer refers to this as "the intuition of the bare two-oneness."[2] Brouwer really does not explain much about the nature of this intuition. Perhaps he thought that what he was saying about it should be immediately clear to everyone. In any event, it seems that the intuition of two-oneness is not itself the idea of a number but serves to generate the numbers. We thus might call this intuition a "number-generator." In speaking of the intuition of two-oneness Brouwer seems to be referring to the ability of the mind to generate a sequence of numbers. He says that this intuition gives rise to the intuition of all finite ordinal numbers, that is the numbers first, second, third, ... up to the idea of the first infinite ordinal.

Arend Heyting has explained the basic intuition in the following way. He assumes that any human being can focus his attention. When we focus our attention we bring into being an object—the object of our attention. Heyting calls such objects "perceptions." Having created one object in this way, he thinks, we can retain this perception in our memory while at the same time we focus our attention on a new object. This object also can be kept in our memory and the process can be repeated indefinitely. This gives rise, he suggests, to the natural numbers, that is to the counting numbers, one, two, three, etc. We find in our minds, for example, five distinct perceptions. We can then abstract from the content of each perception thus producing the idea of the number five—the idea of five units.[3]

Several times in this work we have referred to real number theory the axioms of which we presented in section 4 above. In this chapter we shall refer to the above theory of real numbers as "classical real number theory." The intuitionists have been led to develop their own theory of real numbers due to dissatisfactions with some principles of the classical theory. The intuitionist real number theory differs in significant respects from the classical theory. Some of the differences will become clear as we go on.

In classical set theory and number theory, once the natural numbers exist, the integers (signed numbers) and the rational numbers can readily be constructed providing that we allow that ordered pairs of natural numbers also exist. Identity, addition and multiplication of such pairs is readily definable and shown to have the properties which we usually attributed to the integers and rational numbers.[4] The intuitionists do not give the same account of the theory of rational numbers; however the account that they give is equivalent mathematically. Given the intuitionist point of view, rational numbers must be generated by the basic activity of the human mind. Thus, according to Brouwer, the basic intuition of two-oneness "gives rise immediately to the intuition of the linear continuum, i.e., of the 'between,' which is not exhaustible by the interposition of new units ..."[5] What Brouwer seems to be saying here is that the activity of the mind creates not only a sequence of distinct units but also creates a sequence of subdivisions of any unit. The subdivisions of the first unit yield the rational numbers between zero and one. We can, perhaps, first generate the number $\frac{1}{2}$. But the process of subdividing can be repeated indefinitely, thus generating further rational numbers.

As conceived by Brouwer, the basic intuition generates no facility for creating irrational numbers. The process of sub-division can yield numbers such as $\frac{1}{2}, \frac{1}{4}, \frac{1}{3}, \ldots$ but it cannot yield a number such as the square root of two. If it were possible to obtain the square root of two by a sequence of subdivisions then there would be some natural number N (the number of terms in the sequence of subdivisions) such that the product of N with the square root of two would equal some natural number. (For any rational number $\frac{p}{q}$ there is a natural number N such that the product of n with $\frac{p}{q}$ is a natural number.) There is no natural number N such that the product of N with the square root of two is a natural number.

Due to the work of the mathematicians Cauchy and Dedekind, it is possible to develop classical real number theory as given above by specifying certain sets of rational numbers to be identified as real numbers. One such identification involves identifying real numbers with Dedekind cuts (see section 10 above). If the operations of addition, etc., are defined for the Dedekind cuts, it can then be proved that the set of cuts with the operations in question satisfies the axioms for real number theory. Alternatively, the real numbers can be identified as certain sets of Cauchy sequences of rational numbers. (A sequence of rational numbers R is said to be a Cauchy sequence, if for every number $\in$ greater than zero, there is an integer N such that $|r_i - r_j| < \in$ if both i and j are greater than or equal to N.) Question might be raised as to whether the intuitionists could accept the statements about the real numbers as being true either of Dedekind cuts or of the appropriate sets of Cauchy sequences. The answer to this question is negative.

The definitions of the Dedekind Cuts or of the Cauchy

Sequences do not satisfy Intuitionist criteria for mathematical knowledge. The objects in question cannot be generated from the basic intuition. Objects which can be so generated are said, by the intuitionists, to be constructible. Other mathematicians also speak of constructions of various entities, but they use the term "contructive" in a broader sense. According to the intuitionists, if an object is constructible then one must be able to decide definitely what properties the object has by inspecting his creation. Thus, if there is a property P and one wishes to assert that there exists a set of objects which are all and only the objects possessing P, then it must be possible to decide by inspecting any object whether it has P. For example, if one were to identify the real numbers as Dedekind cuts of rationals then the position of any such real number with respect all other rational numbers should be determined. But there are Dedekind cuts for which this is not true. One such Dedekind cut is Euler's Constant C. Euler's Constant is defined as

$$\lim_{n \to \infty} (\sum_{K=1}^{N} \tfrac{1}{K} - \log n)$$

It is not known whether Euler's Constant is rational. It has not been possible to determine its position with respect to all other rationals. Thus, there are rational numbers which can be constructed such that we do not know whether they are less than C or not and so we could not tell by inspection whether they are members of C (conceived as a Dedekind cut) or not.

One might argue against the intuitionist criteria of existence of numbers, i.e., that the numbers be "constructible", in the following way: Granted that we cannot determine whether certain objects are members of the above Cut C. Nonetheless, for any rational number R, either it is less than C or it is not less than C. Thus, the membership in the set in question is perfectly definite even though we cannot determine what it is. Thus the Cut C should be accepted as a legitimate mathematical object. Heyting, however, anticipated this sort of objection. However, he does not grant its conclusion since he along with Brouwer and other intuitionists rejects the claim that we know that for any rational number R that either R is less than C or it is not less than C. We could know this only if we could construct C and so determine the position of any R with respect to C by inspection.

The critic of intuitionism might argue that we should simply consider the formula by which C is defined. This tells us that C is the limit of a sequence of sums, that is, of the sequence

$$(\tfrac{1}{1} -\log 1), (\tfrac{1}{1}+\tfrac{1}{2} -\log 2), (\tfrac{1}{1}+\tfrac{1}{2}+\tfrac{1}{3} -\log 3), \ldots$$

We can compute this sum after each of the terms in the series. The sequence of terms generated thereby converges toward a limit. So, given any rational number R, either R equals C or we can determine in a finite number of steps whether R is less than or greater than C. In any case the position of R with respect to C is definite. The problem with this argument for the intuitionist is of course that it may turn out that there is some rational number R such that R equals C. Suppose there is such an R, then as we compute the sums in the above sequence we shall never, no matter how many terms of the sequence we consider, reach a point at which we can tell whether R is less than or greater than C since, to reach such a point, we should have to inspect all of the terms of the infinitely long sequence. But, while a human mind can generate a sequence of objects which has no end and can know that no matter how many terms of the sequence have been constructed further terms can yet be constructed, the intuitionists will not agree that the human mind can inspect all of the terms of an infinite sequence. I believe that they would say that this is impossible because in order to inspect all of the terms, the construction of the sequence would have to be completed but, such a sequence can never be completely constructed. For example, if you are constructing the natural numbers in order one by one you can never reach a number about which it would be correct to say "That is the last one."

Heyting has argued against the criticism in question by pointing out that the critic's argument depends on the assumption that we know that the law of exluded middle is true. The critic assumes that we know that after a finite number of steps we shall know either that R is less than C or R is greater than C or we shall not know this after a finite number of steps in which case R equals C. That is, the critic assumes that there is a natural number N such that after N terms of the sequence it is decided either that R is less than C or C is less than R, or there is no such natural number in which case R equals C. But, Heyting argues

the existence of N signifies nothing but the possibility of actually producing a number with the requisite property, and the non-existence of N signifies the possibility of deriving a contradiction from this property. Since we do not know whether or not one of these possibilities exists, we may not assert that *N* either exists or does not exist. In this sense we can say that the law of excluded middle may not be used here.[6]

The intuitionist rejects the law of excluded middle as a valid principle of reasoning because, he argues, if it were accepted as a valid rule of proof, then one could give mathematical proofs of the existence of objects which had not been constructed, that is, which had not been generated from the basic intuition. But, at this point, someone might argue that the intuitionists have conflated the meaning of two different statements. To say of an object that it exists is one thing; to say of an object that it can be constructed is another. Why should we not say of the Dedekind Cut *C*, for example, that this object exists but has not been constructed. We would say of the solar system that it exists even though it has not been constructed. To this suggestion, I think, the intuitionists would reply that with respect to physical objects such as the solar system we have come to know through our experience of the existence of objects which no one has constructed. Thus it is meaningful to speak of objects of that sort (physical objects) which exist but have not been constructed. However, we have no experience of mathematical objects of that sort since our only experience of mathematical objects has been with respect to objects that we have constructed. Thus, in the context of mathematics, the only meaning which "to exist" can have is "to be constructed." If you assert that some non-constructed object exists in mathematics, you are using "exists" in some metaphysical sense which has no place in mathematics.[7]

Given that all mathematical objects must be constructed in the mind, one might think that intuitionists would eschew all statements concerning properties of infinitely large sets of objects. In fact, some intuitionists have taken this position. But, as we have indicated above, some intuitionists have not found this position necessary. Intuitionists of the latter sort can accept claims such as, there are an infinite number of prime numbers, providing that such claims are properly understood. Such claims must not be understood as implying that the individual has constructed in his mind all of the prime numbers and has inspected the entire set and found that it is infinite in number. However, according to Heyting, it may be understood as a statement concerning a hypothetical construction.[8] To say that there are an infinite number of primes may be understood as saying that if one has constructed a natural number *n* then he can construct a larger prime. To prove such a statement, according to Heyting, one must show how the larger prime can be constructed. This construction (of the larger prime) when combined with the construction of the natural number in question would be a proof of the existence of the prime in question. (To construct the prime in question we first construct $n! + 1$ and then factorize this number. Each of its prime factors is greater than the natural number n.) The process of construction must, of course, be a process generated from the basic intuition of two-oneness.

We have argued above that in mathematics some principles are known as a result of a proof and some principles are known without proof. According to the intuitionists some principles are known immediately on inspecting the objects that the mind has created. However, some constructions are very complex and properties of such constructions may not be recognized immediately by someone who has not yet made the construction (or the mathematician may himself forget how he made the construction). In such cases it may be helpful to have some written statements indicating how the construction is to be performed or which objects we ought to inspect (introspect) in order to see that the complex construction has the property in question. Thus, on the intuitionist view, a person may acquire some new mathematical knowledge by studying a proof. However, for the intuitionist there is no fundamental difference between what is known through a proof and what is known without proof, since in both cases the individual acquires knowledge from inspecting an object which he has mentally constructed. Further, according to the intuitionist the following sequence of statements would not be a proof, since the first statement does not reflect any construction that can be performed:

1. Either Fermat's last theorem is true or Fermat's last theorem is not true.
2. If Fermat's last theorem is true then $2 + 2 = 4$.
3. If Fermat's last theorem is not true then $2 + 2 = 5$.
4. Thus, $2 + 2 = 4$ or $2 + 2 = 5$.

38. The connection between the law of excluded middle and the ontological and epistemological views which we have been considering should be thoroughly explored. We shall undertake such exploration in the next two sections. In section 38 we shall consider the law of excluded middle itself and in section 39 we shall consider in more detail the consequences in respect to knowledge of

applied mathematics of the intuitionist principles. We shall take the following as our statement of the law of excluded middle:

(EM) For any statement S, either S is true or the negation of S is true.

Denial of (EM) is tantamount to asserting that

(NEM) For some statement S, it is not the case that S is true and it is not the case that the negation of S is true.

It might be argued that the intuitionist rejection of the law of excluded middle is false since (NEM) is logically inconsistent. This argument might go as follows: To assert NEM is, in effect, to assert that for some statement S, both S is true and S is not true and this is contradictory. However, the intuitionists could (and would) avoid this conclusion in the following way. They would argue that deriving a contradiction from the acceptance of (NEM) entails accepting as a valid principle the principle that

(DN) It is not the case that the negation of S is true is logically equivalent to S is true.

Since the intuitionists have no wish to embrace inconsistencies they willingly give up not only the law of excluded middle (EM) but also the law of double negation (DN). They would argue that (NEM) is not logically inconsistent. To say that it is not the case that S is true is only to say that the mathematician has not been able to construct the objects which would exist if S were true. To say that it is not the case that the negation of S is true is only to say that the mathematician has not been able to deduce a logical contradiction from the assumption that S is true. Surely it is entirely possible both that a person cannot perform the constructions required by some statement S and at the same time that he cannot deduce a contradiction from the assumption of S.

It should be noted that for the intuitionist, to deduce a contradiction from the assumption that S is true is to show that if S were true then one would be able to create an object which would prove a statement that is known to be false. For example, one would show the negation of S by showing that on assuming that S is true one could prove that one is the same as two. The law of contradiction, that is

(LC) It is not the case that some statement S is true and that the negation of S is true is not logically equivalent, in the intuitionist view, to the law of excluded middle. To reject the law of contradiction is to hold that there is some statement S such that one can perform the construction which proves S and that one cannot perform this construction. The intuitionist certainly is not committed to holding that one can and cannot construct the same object.

The rejection of the law of excluded middle and its consequent denial of the validity of argument by reduction to absurdity strikes many people as unreasonable. In mathematics, as practiced by most mathematicians, reduction to absurdity is widely used and, of course, is considered as a valid argument form. Such mathematicians might be led to reject intuitionism simply because of its denial of (EM). Is there any rational way of resolving this conflict? Apparently there is. The motivation of the intuitionist underlying his rejection of (EM) is that through the use of arguments based on this principle, it is possible to "prove" the existence of objects which cannot be constructed. Several philosophers have proposed alternative means of accomplishing this intuitionist objective.

One such suggestion was advanced by Storrs McCall.[9] McCall has suggested that the intuitionists could retain the law of excluded middle provided that they reject the law of bivalence. His formulation of the law of excluded middle is different than that given above and we shall see why below. The law of bivalence is expressed as follows:

(BV) Every statement S is either true or false.

Now, given that for any statement S, S is false if and only if the negation of S is true, we can see that (BV) is equivalent to

Every statement S is either true or the negation of S is true.

That is (BV) is equivalent to (EM) as we have formulated it. However, McCall offers an alternative formulation of the law of excluded middle namely

(EM′) Every statement S is such that either S is the case or S is not the case.

But, what is the difference between (EM′) and (EM)? According to McCall's view, there is a difference between these two principles since McCall rejects the principle (which he attributes to Tarski and Aristotle) that

(T) For any statement S, 'S' is true if and only if S is the case.

Due to his rejection of (T), McCall calls this theory of truth "non-classical." McCall suggests that by denying (T) the mathematician can accept (BV) and reject (EM'). And, McCall suggests, this policy would be more agreeable to mathematicians' intuitions.

However, I doubt that mathematicians will find McCall's suggestion any more palatable than the original policy of the intuitionists. For one thing, the non-classical theory of truth advocated by McCall has other consequences which mathematicians may find disagreeable. One example, noted by McCall, which must be rejected on a non-classical theory of truth is the principle

For any statements S and S', if it is true that either S or S' then either S is true or S' is true.

Even more significant perhaps than consequences such as this is the fact that, according to McCall, arguments by reduction to absurdity are of dubious validity.[10] By taking up McCall's suggestion in place of the intuitionist's suggestion regarding the law of excluded middle, mathematicians would not regain the logical power provided by reduction to absurdity argument forms.

Finally, from an ontological point of view, the non-classical theory of truth advocated by McCall is puzzling. McCall rejects the principle that

(T') 'S' is the case only if 'S' is true (for every S).

In light of the rejection of (T') it is apparently consistent for a person to hold both that it is the case (for example) that snow is white but it is not true that snow is white. But, if I thought someone held both that snow is white and that it is not true that snow is white, I should regard his view as either inconsistent or obscure. I should suggest perhaps that he was confusing "It is true that snow is white" with the proposition that someone knows that snow is white. In any case, I do not find a non-classical theory of truth very plausible.

The argument to which McCall appeals in order to question the validity of argument by reduction to absurdity is an argument which purports to establish the existence of the least natural number possessing the property A, where the property A is identified by the following definition:[11]

"The natural number n has the property A if and only if there are at least $100 - n$ perfect numbers."

However, the question of how many perfect numbers there are is as yet undecided. Thus, we cannot determine which natural number is the smallest natural number that possesses A. Since the existence of the smallest natural number possessing A was established by an argument by reduction to absurdity but since we cannot actually determine which natural number is the smallest natural number which possesses A, McCall thinks that the validity of arguments by reduction to absurdity is questionable. However, we can avoid questioning the validity of reduction to absurdity arguments in general while agreeing that the argument which allegedly establishes the existence of the smallest natural number which possesses A is dubious, if we instead raise question concerning the type of definition used in defining the property A. The definition in question makes reference to a mathematically undecided proposition. What is dubious about the argument in question, if anything is, is the reference to the number of perfect numbers where that number is not known. (A perfect number is defined as a number which is equal to the sum of its proper divisors, e.g., $28 = 1 + 2 + 4 + 7 + 14$.)

The same sort of suggestion can be made with respect to Heyting's argument in which he questioned the assumption of the law of excluded middle. Remember that we spoke of the definition of Euler's constant C and claimed that if C is not rational then for any rational number R there is a natural number N such that after the sum that defines C is computed to N steps then we will know the position of C with respect R. The problem is that it is not known if C is rational and so sets defined with reference to C may not be constructible. Heyting suggests that the following argument which is based on the law of excluded middle is not valid:

If N exists then the set of rationals less than Euler's constant exists.
If N doesn't exist then the set of rationals less than Euler's constant exists.
Thus, the set of rationals less than Euler's constant exists.

However, the problem here can also be traced to the reference to a non-constructible set. If references to sets which are not known to be constructible were eliminated, then the law of excluded middle could not be used to establish the existence of such entities. Perhaps we should eschew use of definitions which make reference to undecided propositions and to propositions which make reference to sets which are not known to be constructible. As Quine has noted, the intuitionists could retain the law of excluded middle and restrict their axioms which assert the existence of sets to axioms which assert only the existence of constructible sets.[12]

The validity of the law of excluded middle was also questioned by Wittgenstein in his work *Remarks on the Foundations of Mathematics.* Some of his criticisms suggest that his view of mathematics was close to that of the intuitionists. Let us discuss some passages of this work in which these criticisms are expressed. In one place he is discussing the question of whether a particular sequence ∅ of numerals ever appears in the decimal expansion of an irrational number such as pi or the square root of two. The question is, is it true that either the sequence ∅ appears in the expansion or that it does not appear. Wittgenstein argues as follows:

> ... men have been trained to put down signs according to certain rules. Now they proceed according to this training and we say that it is a problem whether they will ever write down the pattern ∅ in following the given rule.
>
> But what are you saying if you say that one thing is clear: either one will come on ∅ in the infinite expansion, or one will not?
>
> It seems to me that in saying this you are yourself setting up a rule or postulate.
>
> What if someone were to reply to a question: 'So far there is no such thing as an answer to this question'?
>
> So, e.g., the poet might reply when asked whether the hero of his poem has a sister or not—when, that is, he has not yet decided anything about it.
>
> The question—I want to say—changes its status, when it becomes decidable. For a connexion is made then, which formerly *was not there.*[13]

What is Wittgenstein arguing here? He seems to me to be saying that the sequence of numerals in the decimal expansion of, say pi, does not exist prior to its having been expressed. Thus, the question of whether ∅ exists in the sentence prior to its having been expressed has no answer. It is just like the question with respect to the hero's sister. Thus, it is a mistake to say that either ∅ exists in the sequence or it does not. Someone who says this is using a misleading picture because the picture suggests that the sequence already exists.

The analogy that Wittgenstein makes here with regard to the

completion of a poem does not seem very relevant. I would argue that the question as to how the sequence of numerals in the expansion of pi is to be continued is very unlike the question as to how a poem is to be continued since in expressing pi in decimal form one is following a rule which determines the expansion uniquely. If someone puts down the numeral 6 in the 10^{10} place in the expansion of pi he may be mistaken. Whether he is mistaken could be determined by anyone who carefully computed pi to enough places. This is not the case with regard to the continuation of a poem. One continuation may be better in any number of ways than another continuation of the poem, but no continuation is false and none is true.

Wittgenstein seems to have anticipated this response, for, further on, we find the following comments:

> Good,—then we can say: "It must either reside in the rule for this series that the pattern occurs, or the opposite." But is it like that? "Well, doesn't the rule of expansion *determine* the series competely? And if it does so, if it allows of no ambiguity, then it must implicitly determine *all* questions about the structure of the series." —Here you are thinking of finite series.
> "But surely all members of the series from the 1st up to the 1000th, up to the 10^{10}th and so on, are determined; so surely *all* the members are determined." That is correct if it is supposed to mean that it is not the case that e.g., the so-and-so-myth is *not* determined. But you can see that *that* gives you no information about whether a particular pattern is going to appear in the series (if it has not appeared so far). *And so we can see* that we are using a misleading picture.[14]

Why does the remark that the rule determines the series completely imply that you are thinking of finite series? This does not seem to be the case at all. Note that a finite series may not have been completely expressed. Consider the series which consists of the numerals of the decimal expansion of the square root of two out to the 10^{100}th place. Probably no one has ever completely computed this series. Yet it is finite. If Wittgenstein is suggesting that it is correct to say that either $\emptyset$ appears in this series or it does not appear, then what grounds can he have for denying the applicability of this principle (the law of excluded middle) to the infinite series?

Wittgenstein's statement quoted above implies that the statement of the rule gives you no information concering whether $\emptyset$ appears in the series. But this is not necessarily correct. It may be that the statement of the rule in question would indded give some information which could be used to determine whether $\emptyset$ appears in the series. It may be however that no human being has, as yet, recognized the information in question and been able to utilize that information in the development of a proof which shows either that $\emptyset$ appears in the series or that it does not. And this fact, that no one has as yet been able to utilize the information contained in the statement of the rule to decide whether $\emptyset$ appears in the series does not imply that the continuation is not determined.

The reference to a misleading picture in the last line of the above quotation picks up another strand of Wittgenstein's reasoning. Earlier in the above work he claimed

> When someone sets up the law of excluded middle, he is as it were putting two pictures before us to choose from, and saying that one must correspond to the fact. But what if it is questionable whether the pictures can be applied here?[15]

Then he also argues

> In an arithmetic in which one does not count further than 5 the question what $4+3$ makes doesn't yet make sense. ... That is to say: the question makes *no more* sense than does the law of excluded middle in application to it.[16]

Here, Wittgenstein was arguing perhaps that in an arithmetic in which does not count beyond five it is not correct to say that either $4+3=7$ or that $4+3\neq7$. Wittgenstein may have been thinking that in some cultures the language used does not contain words for numbers greater than 5. In such a language it might indeed make no sense to say that either $4+3=7$ or $4+3\neq7$ since, in such a language the sign '7' might not be a meaningful expression. However, does the fact that such an expression is not meaningful at a particular time or in a particular culture show that a mathematician who uses the principle of excluded middle in a mathematical proof is making an error? I do not think so. It is true, that $4+3=7$ or that $4+3\neq7$, regardless of the fact that this may be meaningless in the language of some culture. For many people in our culture the expression e may be meaningless. Nonetheless it is still true that $e^{i\pi}=-1$ or $e^{i\pi}\neq-1$. Of course, Wittgenstein may not have been thinking about the limitations regarding numerical expressions in other languages. But if he was not thinking of this matter, then what is he thinking when he speaks of an arithmetic in which one does not count further than

5? Conceivably he was thinking of arithmetic modulo five. But if he was then the expression $4+3$ is meaningful. In such an arithmetic presumably $4+3=2$. More than likely Wittgenstein knew about modulo arithmetic and was not referring to it in this quotation. In any case I do not see that he has established the claim that the law of excluded middle is not applicable in mathematical proofs in the ways that most mathematicians wish to use it.

In our discussion of the law of excluded middle, we certainly have not established that it is a valid rule of inference. I do not see how to establish such a claim. However neither the Intuitionists nor Wittgenstein has produced logically compelling reasons for rejecting the principle. If there were no alternative to the Gödelian view regarding mathematical epistemology besides intuitionism and if there were no compelling objections to intuitionism then we might be inclined to accept the intuitionist strictures with regard to the non-existence of complete infinite sets and non-constructible sets. Such an ontology might be quite plausible if one regarded mathematical objects as having been created by the mathematician or at least by a number of mathematicians working cooperatively. However, we have not yet considered some serious objections to the intuitionist theory and there are alternative epistemologies to both intuitionism and Gödelianism.

39. *Intuitionist Theory of the Continuum*

As noted above the intuitionist account of our knowledge of principles of natural numbers, integers and rational numbers rests on our ability to create and mentally inspect mental objects. However, as noted in our discussion of Euler's constant, classical real number theory provides for the existence of objects which cannot be constructed. It is therefore unacceptable to the intuitionist. Yet, as noted earlier, the rational numbers are not complete. This has profound consequences in the theory of calculus and through calculus to physical theories. The development of calculus presupposes classical real number theory as even a brief introduction to the study of sequences and of such properties of sequences as convergence and uniform convergence makes clear. Similar remarks apply to the study of functions with respect to the property of continuity. In the application of mathematics in physical theories many physical processes are represented as functions or as sequences. For example, in studies of the motion of objects distance or velocity are represented as functions. If we agree that we have much mathematical knowledge which is applied in natural sciences and daily life and that such knowledge is not included in the mathematical theory of rational numbers (and certainly not within the theory of integers or natural numbers) then intuitionist theories would be seriously incomplete unless they could provide as satisfactory explanation as to how such knowledge is obtained. To remedy this defect intuitionist mathematicians such as Brouwer, Heyting and others have tried to develop an intuitionist theory of real numbers or of the continuum. Since this book is not a mathematical work we shall not present this theory in detail. We shall merely present some of the basic ideas of the intuitionist theory as presented in Heyting's work *Intuitionism, an Introduction.*[17] This will serve for some brief comparisons of the intuitionist real number theory with classical theory of real numbers. We shall note some points in which the key concepts of intuitionist real number theory are defined differently than they are in the classical theory and we shall note some differences in the mathematical knowledge that is included in the two theories.

In section 37 above we presented the definition of a Cauchy sequence. We have assumed that the reader has at least an intuitive notion of the concept of a sequence of numbers and so have not tried to explain this notion or to give a rigorous definition of sequence. Roughly the idea of a Cauchy sequence is that in a Cauchy sequence of rational numbers, for any number $\frac{1}{k}$ (where k is a natural number) there is some term n of the sequence (it may be very far from the initial term of the sequence) such that the difference between any two terms of the sequence that follow the nth term is less than $\frac{1}{k}$. A sequence is said to be Cauchy in the classical theory if there always is some term (an nth term) with the requisite property. But it is possible that a sequence is a Cauchy sequence and yet we cannot determine which term is the requisite n^{th} term with respect to a given number $\frac{1}{k}$. That is, with respect to a given number $\frac{1}{k}$ we may not be able to calculate the term beyond which the difference between any two terms is $\frac{1}{k}$. This is unsatisfactory to the intuitionist. He holds that if we assert that existence of such an nth term then we must be able to calculate the n in question.

However, the intuitionist has been able to develop a definition of Cauchy sequence which closely parallels the classical definition. It is given by Heyting as follows:

A sequence $\{a_n\}$ of rational numbers is called a Cauchy sequence, if for every natural number k we can find a natural number $n = n(k)$ such that $|a_{n+p} - a_n| < \frac{1}{k}$ for every natural number p.

Note that the intuitionist does not simply assert that n exists but specifies that n can be found.

In the set theories such as *Principia Mathematica*, classical real number theory can be derived by defining real numbers as Cauchy sequences of rational numbers. The intuitionist theory of real numbers parallels this development. However, not only is the notion of a Cauchy sequence defined in a more restrictive fashion. The basic notion of a set is more restrictive. Basically, sets in intuitionist theory have to be built up from the objects which can be constructed. Intuitionist sets are referred to as *species*. In order for an object to be allowed as a member of a species S, it must be defined independently of S. (Thus the intuitionists would agree with Poincaré in ruling out impredicative specifications of sets.) Further, a species is a set of objects defined by a characteristic property of its elements.[18] Of course the property, by reference to which a species is defined, must satisfy intuitionist scruples. For example, to define a species as a set of classical Cauchy sequences would be unsatisfactory. We can now proceed to the intuitionist definition of real numbers.

A Cauchy sequence of rational numbers is said to be a *real number generator*. Two number-generators a_n and b_n are identical if $a_n = b_n$ for each n. Two number generators a_n and b_n are said to *coincide* if for every k we can find $n = n(k)$ such that $|a_{n+p} - b_{n+p}| < \frac{1}{k}$ for every p. (In other words two number generators coincide providing that as you get far enough out on the two sequences involved the corresponding terms differ by less than $\frac{1}{k}$.) The relation of coincidence is an equivalence relation. That is, it can be shown to be reflexive, symmetrical and transitive and so it can be used to define equivalence classes. Real numbers are defined as equivalence classes (or rather species) of real number generators. That is, a species of real-number generators which coincide with a given real-number generator is said to be a *real number*. The species of real numbers is said to be the continuum. (The continuum in classical real number theory is the set of all real numbers also.) Definitions of the operations of addition and multiplication for real numbers can be given which closely parallel the definitions of these operations in classical real number theory and so the arithmetic of real numbers can be obtain in an intuitionist theory. Further, definitions can be given for convergence of sequences, continuity of functions, etc. Heyting gives as the definition of positive convergence the following:

> A sequence a_n of real number-generators is (positively) convergent to the limit a, if given any natural number k, a natural number n can be found such that for every natural number $p |a - a_{n+p}| < 2^{-k}$

Some interesting and powerful theorems can be proved in the intuitionistic theory of real numbers which cannot be proved in the classical theory, for example, on the intuitionistic theory it can be proved that a real-valued function $f(x)$ which is defined everywhere on a closed interval of the continuum is uniformly continuous on that internval.[19] Certain other theorems of classical analysis remain true in the intuitionist theory. For example, in both theories it can be proved that a sequence is convergent only if it is Cauchy. However, other basic principles of analysis cannot be proved in the intuitionist theory of real numbers although they can be proved in the classical theory of real numbers. For example, in the intuitionist theory it cannot be proved that every bounded monotone sequence is convergent. Further, the Bolzano–Weierstrass theorem cannot be proved in the intuitionist theory (but can be proved in the classical theory). This theorem states that for any bounded infinite subset of the real numbers there is a point of accumulation. The problem for the intuitionist version of this principle is that it cannot be proved that the point of accumulation can always be found. Further, the trichotomy law for real numbers cannot be proved on the intuitionist theory.[20] And it cannot be proved that there is a zero for every continuous real function $f(x)$ with $f(a) < 0$ and $f(b) > 0$.[21] The failure of intuitionist mathematicians to establish intuitionistic analogues of these important theorems of classical analysis is significant. These theorems underly applications of mathematical principles in the natural sciences. The intuitionist theory thus fails to provide a satisfactory account of the origin or basis of our mathematical knowledge. Let us elaborate on this point.

We have just claimed that the intuitionist theory does not provide a sufficient basis for all of our mathematical knowledge. But how can we substantiate such a claim? That is, how can we claim that (1) there are mathematical propositions which we know and which are such that (2) these propositions cannot be proved on an intuitionistic basis? Our answer to this is that there are propositions such as those given in the last paragraph which are not provable intuitionistically. But what justifies our claim that these propositions are known? In answer to this we reply as follows: Some of the propositions in question are essential parts of the mathematical development of theories in the empirical sciences which have been widely corroborated by observational evidence. Since the mathematical principles are parts of theories which have been corroborated or confirmed empirically, the mathematical principles have been confirmed also. For example, a mathematical procedure which is widely applied in solving problems in calculus is implicit differentiation. This technique is applied in cases in which one has an equation of the form $f(x,y)=0$ in which it is not the case that y is expressed as a function of x nor is x expressed as a function of y. In this situation it may be impossible to express y as a function of x or to express x as a function of y. Still we may be able to differentiate the terms on both sides of the equation. This is the process of implicit differentiation. In effect, it treats the equation as defining y as a function of x implicitly. The theorem which expresses the conditions under which implicit differentiation is justified is often called an implicit function theorem. The proof of this theorem depends on the claim that there is a zero for every continuous real function $f(x)$ with $f(a)<0$ and $f(b)>0$. Thus, the implicit function theorems cannot be proved intuitionistically. Again, failure of this theorem (the theorem that there is a zero for every continuous real function, etc.) implies that the fundamental theorem of calculus cannot be proved. (The fundamental theorem of the calculus is the result which expresses the fact that anti-derivatives can be used in evaluating definite integrals.) Through applications to practical problems in science and engineering, the implicit function theorems and the fundamental theorem of calculus are highly confirmed and so are the principles which are necessary for proof of these theorems. It is on this basis that I say that we have knowledge that, for example, there is a zero for every continuous real function $f(x)$ with $f(a)<0$ and $f(b)>0$. Similar remarks apply to the other examples we have given of principles of real number theory which intuitionists cannot prove.[22]

   Several responses to this criticism may be imagined on the part of the intuitionist. One possibility is to simply admit that intuitionism does not give an explanation of the nature of our mathematical knowledge. Heyting comes close to suggesting this in the following statement. He admits that intuitionistic mathematics is of very little use in physics and then says:

But is usefulness really the only measure of value? It is easy to mention a score of valuable activities which in no way support science, such as the arts, sports and light entertainment. We claim for intuitionism a value of this sort ...[23]

Of course, Heyting may not have meant to imply that since intuitionist mathematics is not useful, it does not give a complete account of our mathematical knowledge. That is, Heyting may have held that the usefulness of mathematical principles has no bearing on the question of their truth. He might agree with Vaihinger that classical real number theory is useful though false while holding that intuitionist real number theory is not particularly useful but is nonetheless true. This is a position which we have rejected above (section 31). The derivation of experimental consequences from mathematical theories requires the use of mathematical principles such as those stated in classical real number theory. If observation of the empirical consequences corroborates or confirms the theory then such observation confirms the mathematical principles also.

Another reply for the intuitionist to make is that mathematical knowledge is perfectly exact and certain. This seems to be the view of Brouwer. For example he stated in the above cited paper Brouwer refers to the "apodictic exactness" of the axioms of arithmetic.[24] By this he seems to mean that arithmetical concepts are perfectly exact and further that those arithmetical propositions which are known to be true are known with certainty. His view seems to be the same with regard to other mathematical principles also. Then, following the rationalist philosophers and Kant, Brouwer held that propositions based on sensory evidence do not contain exact concepts and such propositions cannot be known with certainty to be true. Clearly, if we hold that mathematical knowledge is certain but that knowledge based on sensory evidence is not certain then we must conclude that mathematical knowledge does not rest on sensory evidence. In the following section we shall further evaluate Brouwer's position. We shall argue that even if our mathematical knowledge is certain, the intuitionist explanation concerning how such knowledge exists is unsatisfactory.

40.   According to the intuitionists (intuitionist) mathematics is absolutely certain because it rests on direct awareness of mental constructions. Thoughts which are correctly describable as being a direct awareness of mental constructions are necessarily true. Their truth is known immediately. They are, in a word, self-evident.

Proofs which satisfy intuitionist standards must preserve self-evidence.

An obvious sort of question to ask with respect to such a theory is whether our awareness of our mental constructions is necessarily true. That is, if we grant that mathematical objects are constructed in our mind and if we grant that we are aware of objects which are in our mind, still, is it not possible that the objects do not have the properties which we think they have. Many philosophers would challenge the intuitionists on this point. For example, Karl Popper has claimed that there are no authoritative sources of knowledge.[25] By this statement Popper meant to deny that any experience yields thoughts or statements which are necessarily true. Stephan Körner has developed this criticism in detail. Körner argues that, if any thoughts are self-evident then any person who has those thoughts while undergoing the experience to which they refer must agree both that the thought is true and that it is self-evident. If there is any conflict then the thought is not self-evident. Does such conflict occur?

Heyting has reported a conflict between Brouwer and another intuitionist mathematician named Griss. Brouwer accepted as self-evident the claim that "a square circle cannot exist." But Griss did not accept this claim as self-evident. Apparently Griss thought that we cannot have a clear idea of objects that do not exist. This leads him to completely eschew negative propositions in mathematics. Negative propositions are essentially unclear and thus non-mathematical from the point of view of Griss.[26] If we accept Körner's claim in regard to self-evident propositions we apparently must conclude at least that no negative propositions are self-evident.

Heyting tried to avoid the force of this criticism by suggesting that conflicts amongst intuitionist mathematicians are due to misunderstanding. According to the intuitionist, it is thoughts which are self-evident. But conflicts amongst intuitionists are evident only when the thoughts are expressed. Unfortunately the language in which one expresses his thoughts is never a perfect representation of the thought. Thus, conflicts can arise between intuitionists because intuitionists, like all of us, are never speaking perfectly precisely. Heyting said

... my mathematical thoughts belong to my individual intellectual life and are confined to my personal mind, as is the case for other thoughts as well. We are generally convinced that other people ... can understand us when we express our thoughts in words, but we also know that we are never quite sure of being faultlessly understood.[27]

He also said that "As intuitionists speak a non-formalized language, slight divergences of opinion between them can be expected."[28] Thus, according to Heyting while there are conflicts amongst intuitionists with regard to what they say, this does not prove that there are conflicts with regard to what they think. What Körner's argument shows is only that the *reports* of intuitionists are not self-evident.

Heyting's reply to Körner's objection is not satisfactory. To see this consider the fact that the intuitionist's theory is intended as an explanation of the purported fact that laws of mathematics are absolutely certain. Brouwer, in the above cited paper, refers to the "unassailable" laws of arithmetic and geometry.[29] Presumably, in referring to the laws of arithmetic and geometry he was referring to statements of such principles in commonly understood languages. He was not claiming that while people have thoughts of arithmetical principles which are necessarily true, their statements of such thoughts are, to some degree, doubtful. Thus, Heyting's reply to Körner's objection amounts, in effect, to denying the purported fact which Brouwer intended to explain. If mathematical statements possessed "apodictic certainty" then Heyting's reply undermines the ability of intuitionism to explain the apodictic certainty of mathematics.

To my knowledge intuitionists have not rejected Körner's claim in regard to the self-evidence of mathematical statements. Körner said

If two reports about the same intersubjective experience, both linguistically correct, are incompatible, then the experience cannot be self-evident, whatever else "self-evidence" may mean. For since a linguistically correct report of a self-evident experience is according to the theory necessarily true, and since two linguistically correct reports which are incompatible cannot both be true, the reported experience cannot be self-evident.[30]

However, even though the intuitionists have not rejected this contention that self-evident reports cannot logically conflict we may ask if it would be reasonable to defend their position by rejecting this contention. That is, could we allow that there can be conflicts amongst reports of the basic principles of mathematics but that the existence of such conflicts does not imply that the reports are not self-evident? This reply does not seem plausible. The very fact that there is incompatibility amongst sincerely expressed reports is a valid reason for doubting the reports. If there is a valid reason for doubting the reports then the reports are not necessarily true and so are not self-evident.

The existence of conflicts amongst allegedly self-evident intuitions goes even further than we have yet indicated. Brouwer and other intuitionists accept as true all of the classical principles of arithmetic concerning the natural numbers, integers and rational numbers. But some intuitionist mathematicians would even reject parts of arithmetic. While Brouwer and Heyting accepted the principle of induction, for example, the mathematician Esenin-Volpin does not.[31] Thus we must even deny that the fundamental principles of arithmetic (Peano postulates) are self-evident. But, if these principles are not self-evident then it seems very likely that no principles are self-evident.

The intuitionist view of mathematical statements suggested by Heyting's remarks above is open to further objections. No one would deny that mathematicians sometimes make false statements. Even an intuitionist mathematician could, it is allowed, make an error. If a mathematician makes an error then presumably his erroneous statement can be contradicted by statements of other mathematicians. But, on the intuitionist theory concerning mathematical statements, such contradiction cannot occur. An argument used by G.E. Moore in his critique of certain subjectivist theories of ethics can be applied here. In order for one mathematician's statement to contradict the statement of another, they both must be talking about the same objects. However, if every mathematician speaks only about objects contained in his own mind and if each person's mind is distinct from all other minds, then it is impossible for logical contradictions to occur. If one mathematician says that a square circle exists and another says that there are no square circles, the intuitionist cannot conclude that one of these statements must be mistaken. For, on the intuitionist view, each mathematician is referring to entities that exist in his own private mind. Since this conclusion is absurd we must reject the intuitionist view concerning the nature of mathematical statements.

Of course, one could try to defend intuitionism against this criticism by suggesting that when a mathematician says that some object exists he means that anyone could construct an object of the specified sort, while if someone denies that such an object exists he means that no one could construct such an object. Then the statements of the two mathematicians do logically conflict.

However, if the defender of intutionism makes this move then it is no longer true that the statements of a mathematician refer only to objects in his own mind. Under the defense here suggested, mathematicians make statements about what can be constructed in anyone's mind. But, it seems quite unlikely that statements about what can be constructed in anyone's mind could be self-evident.

It appears in virtue of the existence of disagreements amongst mathematicians that we must conclude that there are no self-evident principles of the sort presupposed by Brouwer. Efforts to avoid the force of this objection by suggesting that the conflicts are due to the imperfections of language are unacceptable since they deny the very datum that intuitionism takes for granted. Further, the intuitionist theory that mathematicians' language is to be understood as referring only to what is in the language user's mind is incompatible with the view that logical conflicts between mathematicians are possible.

41.   What I want to do in this section is to reflect somewhat more concerning the topic of intuition in mathematics. The topic is of some importance to mathematicians, especially in connection with their thought concerning mathematics education. I am sure that I have heard some mathematicians say that one of the aims of mathematics education ought to be to develop the student's mathematical intuition. I have also heard it claimed that one must have sufficient intuition to cope with advanced courses in mathematics. Yet, in this work so far we have criticized two theories which relied on mathematical intuition as a source of knowledge. We have argued essentially that intuition either in the sense of the term required by Gödel's theory or in the sense of the term required by Brouwer's theory does not exist. We do not have non-sensory awareness of externally existing non-physical objects. Nor do we have introspective experiences which provide infallible guarantees of the truth of mathematical propositions. In light of these criticisms we must interpret modern-day mathematicians who speak of developing a student's intuition as not meaning that they are trying to develop the student's ability to become aware of non-physical objects in a non-sensory way. Similarly, we do not understand such mathematicians to mean that students are not ready for advanced courses because they are not yet ready to introspect infallibly. What then do mathematicians have in mind when they speak of intuition?

I suggest that what they have in mind primarily is the ability of a person to understand abstract concepts. When a mathematician says that a particular student does not have sufficient intuition to be able to cope with a particular course, one of the things he may mean is that the student will not be able to understand the abstract concepts involved. To understand a concept is either to know of which objects it is true or to know under what conditions it is applicable to (or true of) objects. In characterizing a concept as abstract I mean to contrast that concept both with concepts whose applicability is closely tied to sensory observation and with concepts which are relatively narrow in scope of application. If a mathematician claims that a student will not be able to cope with a course because of his inability to cope with abstract concepts, he is saying that the student will not be able to apply the concept correctly to the objects being studied because that student is able to operate correctly only with concepts that are either very closely tied to sensory observation or relatively narrow in scope of application. (Let us call concepts which have a narrow scope of application concrete. Some people are only able to understand concepts which are concrete and empirical.)

In elementary mathematics courses, such as one studies in elementary or high school, the principles one studies are in many cases closely tied to the senses for their application. This is clearly true in the case of elementary arithmetic. The natural numbers are counting numbers. One can correctly apply the concept six or of other counting numbers because one knows what observations are relevant for decisions as to whether the concept applies. Even as one begins to generalize the concept of number by the addition of integers and rational numbers, one's experience with the concepts involved is still closely connected to the senses. And furthermore, the generalized concept of number is still relatively narrow in scope as compared to concepts studied in advanced university courses, e.g., concepts such as that of a ring, or a topological space, or a category. A mathematician who thinks of mathematics courses as developing a student's intuition may think that in the sequence of courses from elementary to advanced there is a gradual development from concrete empirical concepts to abstract concepts. By following such a course of study the student's ability to understand abstract concepts is presumed to increase.

Intuition, as here understoood, can be considered as a kind of knowledge. A student whose intuition has been sufficiently developed knows how to understand abstract concepts. He can understand statements in which such concepts play an essential role. Further, intuition of this sort is an essential prerequisite for the acquisition of knowledge of abstract mathematical principles. If a person does not understand such principles then he certainly cannot come to know that they are true. For example, if he does not understand such concepts then he will not understand mathematical proofs of principles involving such concepts. Nonetheless in speaking of intuition as we have been, we have not been referring to a faculty whose use yields knowledge. Nor is such intuition an experience which yields, when it occurs, self-evident knowledge. It is not a kind of evidence to which we may appeal if called upon to try to show why we believe that a statement is true. However, there is another sense of the term *intuition*. In this second sense intuitive knowledge is knowledge that a person has acquired without having been exposed to a proof. Is there any intuitive knowledge of this second sort?

Our characterization of intuition of the second kind as a kind of knowledge which a person acquired without having been exposed to a proof is too general. Clearly, if we were to define the term *intuition* in this way then we would have to class knowledge gained on the basis of sensory observations as intuitive knowledge. I think it would be more useful to exclude knowledge based on sensory evidence from the class of intuitive knowledge. What we are interested in is whether there are occasions when we would say that a person has come to know a proposition *P* when he has not seen or heard a proof that *P* and when he has not directly confirmed a statement that *P* by deducing observational consequences from it (and further *P* is not an inductive or low-level empirical generalization). Surely, we must allow that there are occasions when people come to know mathematical propositions *P* in this intuitive way. To see this consider that sometimes mathematicians and others often see how to prove mathematical propositions. If a person discovers for himself (and perhaps for all of mankind) how to prove a proposition *P* then presumably he came to know *P not* because he first became acquainted with a proof that *P*. Rather, he came to know *P* and either subsequently or simultaneously discovered a way of proving that *P*.

Discovery of a proof need not be regarded as a mysterious process in which a person is caused to perceive non-physical objects. Albeit it is a process which is poorly understood. It sometimes involves creation of new concepts. It always involves finding logical relationships among statements of which one was formerly unaware. It may occur nearly instantaneously. Or it may be the result of tedious work. Indeed, one may compare the discovery of a proof with the discovery of a way of solving some practical problem, say in carpentry. We do not postulate forms of non-sensory perception in this case. Consider the carpenter faced with a problem which he has not solved before. He can perceive the wood and tools with which he works. But since, by hypothesis, he has never solved this problem before, he cannot have observed the way to solve it. Why don't we attribute to him a non-sensory perception of ideal pieces of wood being arranged by ideal tools in just the way which solves the problem? Clearly, we do not think that such a supposition is necessary. His discovery of a solution may be attributed to his having perceived an analogy between the present problem and other problems which he has seen solved in the past. Or it may be suggested that on the basis of his past knowledge he was able to make an informed guess as to how the immediate problem could be solved and then on reflection saw that what he had guessed would work. I see no reason why similar remarks cannot be applied in the case of mathematical discovery. Discovery of a proof might involve the making of an informed guess (a guess that reflects prior knowledge) followed by a process of reflection in which one works out the logical implications of one's guess.

If we allow that there is intuitive knowledge involved in cases in which a person discovers a proof, should we also be willing to allow that sometimes a student may have intuitive knowledge of a provable mathematical principle even though he cannot formulate a proof of that theorem? Teachers sometimes speak of cases in which a student comes to accept a proposition P without have seen or heard a proof of P but in which the student cannot formulate a proof of P. In such cases in which P is true, should we say that the student knows P. My inclination, in light of philosophical tradition, is to say that since the student cannot formulate a proof of P that he does not know P. The student merely believes P. Yet, I am not persuaded that describing the situation in this way is completely adequate. Couldn't it be that the student had become aware of a way of utilizing prior knowledge which does in fact yield a proof of P but that he was not able to hold on to this awareness. Perhaps, before he could express what he thought in words his awareness of the proof disappeared. Shouldn't we say in such a case that the student knows P but he doesn't know a proof that P. Consider, Fermat's last theorem, perhaps Fermat had a brief insight in which he became aware of a proof of this theorem but never got to write it down. If we assume that the theorem is true, then it might be correct to say that Fermat knew the theorem though he had never formulated a proof of it.

In any case, I think we should grant that there is intuitive knowledge of this second sort. That is to say, a person might undergo an experience of a rather complex and poorly understood sort which we can refer to as a process of acquiring knowledge through intuition. However, intuitive knowledge of this sort does not appear to be a way of solving the basic problem regarding mathematical epistemology. As we have described this intuitive knowledge it is knowledge of a proof or knowledge of a theorem which can be proved (if we allow the case in the person comes to know a theorem but cannot formulate a proof). But this intuition does not appear to be a way of coming to know the basic axioms. It is rather a way of using principles that one knows to come to know something new.

42.   Conclusion: Our discussion of mathematical knowledge has been aimed at determining how knowledge of basic mathematical principles can be acquired. In this light we have considered Gödel's view and also intuitionism and have found neither of these views to be adequate. Gödel and the intuitionists suggest that such knowledge claims rest on non-empirical intuitions. We have next to consider the possibility that there are empirical grounds for the basic propositions of mathematics. This is a task that we shall undertake. However, on some views the concept of empirical knowledge is inconsistent. On such views, if a person knows a proposition P then that person is certain that P is true. But, if the evidence that one has for the truth of P is empirical then one cannot be certain that P. Thus, such views conclude that empirical knowledge does not exist. I believe such views to be mistaken. In order to lay the groundwork for a theory of empirical mathematical knowledge I shall in the next chapter try to show that we are not forced to choose between claiming that our knowledge is absolutely certain and claiming that we have no knowledge at all. A third alternative is a view according to which our knowledge is fallible.

NOTES

1. Brouwer, (5) p. 69.
2. *Ibid.*, p. 69.

3. Heyting, (32) p. 80 ff.

4. For an account of this development see Waismann, (90).

5. Brouwer, (5) p. 69.

6. Heyting, (29).

7. Heyting, (32) p. 81.

8. Heyting, (32) p. 81.

9. McCall, (53) pp. 83–88.

10. *Ibid.*, p. 88.

11. *Ibid.*, pp. 87. McCall credits the argument to E.W. Beth in his book *Mathematical Thought* (Dordrecht, 1965) p. 82.

12. Quine, (74) p. 88.

13. Wittgenstein, (91) p. 138e.

14. *Ibid.*, p. 139e.

15. *Ibid.*, 139e.

16. *Ibid.*, 140e.

17. See also Troelstra, (88).

18. Heyting, (31) p. 37.

19. *Ibid.*, p. 47.

20. *Ibid.*, p. 122.

21. Fraenkel, (18) p. 268.

22. I am indebted to my wife for calling my attention to these examples which illustrate the importance of the theorems which intuitionists cannot prove in mathematical theorems of wide application.

23. Heyting, (30) p. 62.

24. Brouwer, (5) p. 67.

25. Popper, (63) p. 132.

26. Heyting, (30) p. 63.

27. *Ibid.*, p. 61.

28. *Ibid.*, p. 63.

29. Brouwer, (5) p. 67.

30. Körner, (38) p. 136.

31. For further discussion of this point see Fraenkel, (18) p. 251.

Chapter Seven

Mathematical Knowledge as Empirical

43.   When we think of the view that mathematical knowledge rests on sensory observations, we most often think of John Stuart Mill. Mill is the most well-known defender of this position, though the view has had other defenders in recent years.[1] Mill's position on this matter was subjected to such strong criticisms in this century that many philosophers, I am sure, consider the view that mathematical knowledge rests on sensory evidence to be obviously false. However, it is my opinion, following on the work of Quine and Putnam, that the view that mathematical knowledge rests on sensory observation is essentially correct.[2] It will thus be useful to criticize Mill's critics and show that their efforts to draw the conclusion that mathematical knowledge was entirely *a priori* were unsuccessful.

To say that the view that mathematical knowledge rests on sensory observation is Mill's view is not, strictly speaking, correct. It is a generic theory of which Mill's view is a specific version. In fact, we do not wish to show that the specific version of this theory which Mill accepted is correct. We believe that this is not the case. Mill was mistaken, however, not because he thought that mathematical knowledge is supported by empirical evidence but because of the specific way in which he thought that sensory evidence supports mathematical conclusions. To make this point clear let us distinguish empirical generalizations from empirically supported theoretical propositions. An empirical generalization is a statement of the form "All F's are G's" or "M/N F's are G's" where the F's and G's are observable objects, properties or relations. A statement S is an empirical generalization providing that S has one of the forms in question and further is inferred to be the case after having observed a sample of F's which were G's. Saying that a statement is an empirical generalization is saying that one has come to know it through an inductive inference of the above sort—an inference of a general principle from observations of particular cases.

An empirically supported theoretical proposition is not an

empirical generalization in the above sense. That is, it is not confirmed or warranted simply on the basis of an inductive inference of the form envisioned by Mill. A theoretical proposition is empirically supported if the evidence by reference to which it is confirmed via the so-called hypothetico-deductive method. That is to say, a proposition S is empirically supported in this way if S, taken in conjunction with other propositions, entails statements which report observations and if such observation statements are verified. Consider for example, the statement that *homo sapiens* evolved from *homo erectus*. This statement might be confirmed empirically if taken in conjunction with other statements it entails observation reports (and these observation reports are not entailed by the other statements alone) and these observation reports are verified. With reference to this example the observation reports would presumably be statements about fossil remains. The other statements used to deduce the statements about fossil remains would include statements concerning what happened to the skeletons of *homo erectus* individuals and *homo sapiens* individuals upon the death of such individuals.

That Mill held that arithmetical propositions are empirical generalizations seems clear from the following quotation:

> It is a truth known to us by early and constant experience, an inductive truth, and such truths are the foundation of the science of number. The fundamental truths of that science all rest on the evidence of sense; they are proved by showing to our eyes and our fingers that any given number of objects—ten balls, for example—may by separation and rearrangement exhibit to our senses all the different sets of numbers the sums of which are equal to ten. ... The science of number is thus no exception to the conclusion we previously arrived at that the processes even of deductive sciences are altogether inductive and that their first principles are generalizations from experience.[3]

Surely Mill is saying here that we have observed that five fingers plus five fingers is ten fingers and that whenever we have observed two distinct groups of five objects each, we have observed that there were ten objects in the combined collection of the objects. Further, Mill is saying that our knowledge that $5 + 5 = 10$ rests on just these observations. In defending Mill's view in the following pages we shall not want to defend the view that arithmetical propositions are empirical generalizations.

44.   Although the logical positivists were, like Mill, empiricists, they rejected Mill's contention that the propositions of arithmetic

are empirical generalizations. Their arguments however support an even stronger conclusion. Their arguments, if sound, would establish that mathematical propositions are not supported by sensory evidence at all. The arguments they used in criticism of Mill are extremely simple. Essentially the same argument is found in the work of both A.J. Ayer and Carl Hempel. It is roughly as follows: If a proposition is an empirical hypothesis then it is at least theoretically disconfirmable by observations. That is, if a hypothesis is an empirical hypothesis then "it is possible to indicate what kind of evidence, if actually encountered, would disconfirm the hypothesis."[4] Consider then some arithmetical propositions such as "$2 \times 5 = 10$." What evidence would disconfirm this proposition? It is most natural to suppose that the occurrence of two non-overlapping groups of five observable objects each in which the sum of all the objects in the two groups was found by counting not to equal ten would disconfirm the proposition. However, according to the positivists, should we actually encounter this evidence it would not disconfirm the proposition. Should we actually encounter this situation we would suppose that there were not "five pairs of objects to start with, or that one of the objects had been taken away while I was counting, or that two of them had coalesced, or that I had counted wrongly ... The one explanation which would in no circumstances be adopted is that ten is not always the product of two and five."[5] Since, then, even in this sort of case there is no disconfirming empirical evidence, there is simply no disconfirming evidence for arithmetical propositions and so arithmetical propositions are not empirical hypotheses of any sort (neither empirical generalizations nor empirically supported theoretical propositions). This is the full significance of the positivist critique of Mill. They argued this way in support of their view, which we have discussed above, that mathematical propositions are analytic or true by definition.

Actually, in Mill's writings one can find the positivist view of mathematical propositions clearly anticipated. Mill distinguished between arithmetic as a science of "pure number" and arithmetic as a science of quantity. In the latter science the propositions are empirical generalizations. In the former science, one makes the assumption that all units are perfectly equal. Propositions of this science may be exact and certain. This distinction between the science of pure number and the science of empirically given quantities suggests that Mill was applying to arithmetic a view which the positivists applied to geometry but not to arithmetic.[6]

But Mill would not have agreed, I do not think, with the positivist view that the propositions of pure number are analytic. Rather, his view with respect to this science is probably closer to that of Vaihinger. For, Mill held, pure numbers do not exist. He said "there are no such things as numbers in the abstract."[7] The positivists would presumably have held that it is meaningless to make such a claim as this.

Is the positivist criticism of Mill correct? We think not. It appears that they did not consider the possibility of disconfirming evidence seriously enough. In the example above in which the product of two and five is observed not to be ten, Ayer and Hempel introduce various auxiliary hypotheses in the effort to explain why in that particular case the product did not equal ten. But these auxiliary hypotheses can all be tested. We can re-observe the two groups and see if each one still has five objects. We can re-count the objects in the two groups and re-count the total number of objects in both groups. We can take steps to ensure that none are removed while counting. We can take steps to prevent the objects from coalescing or mutliplying (if they are the sort of objects that do that sort of thing). Suppose that we had done everything that we could to maintain the constancy of the groups had checked and verified the correctness of our counting and still found that the groups counted separately had five objects each but that when they were counted together they did not have ten objects. What would we do in that case? It seems to me that the answer is not clear. We could consider the possibilities of mass hypnosis or some other form of mass illusion such as a deceitful demon. We could also consider the possibility that our arithmetic couldn't be applied to these objects, that is to say, that our arithmetic principles would be false. To insist that even in this case we would still refuse to consider that the arithmetic principles are false is unwarranted.

No doubt, should such unexpected results of counting occur, we would feel quite confused. Some investigators would continue to retain the normal arithmetic and search for explanations of this strange phenomenon consistent with that arithmetic. But, other investigators might try applying alternative arithmetics. They might consider arithmetics based on different sets of numbers or arithmetics in which the numbers didn't have a linear order.[8]

45.　　Gottlob Frege, in his work *The Foundations of Arithmetic*, subjected Mill's theory concerning the nature of arithmetic truths to extensive criticism. He hoped to show that Mill's view that the laws of arithmetic are supported by sensory evidence could not be true. Since we defend the view that the principles of mathematics including arithmetic are supported by sensory evidence, it is incumbent upon us to consider Frege's criticisms. His first criticism of Mill concerns the definitions of the numerals such as 3,4, ... Mill allowed that $2 =_{df.} 1 + 1$, $3 =_{df.} 2 + 1$, etc. However, Mill thought that these definitions do more than merely fix the sense of the symbols 2,3, etc. In addition to fixing the sense of the symbols these definitions implicitly contain statements asserting the existence of objects of which the definition is true. That is, there are collections of which the definition of '2' is true, etc. Against this suggestion Frege argues:[9]

> He informs us, in fact, that these definitions are not definitions in the logical sense; not only do they fix the meaning of a term, but they also assert along with it an observed matter of fact. But what in the world can be the observed fact, or the physical fact (to use another of Mill's expressions) which is asserted in the definition of the number 777864? Of all the whole wealth of physical facts in his apocalypse, Mill names for us only a solitary one, the one which he holds is asserted in the definition of the number 3. It consists, according to him, in this, that collections of objects exist, which while they impress the senses thus, $_o^o{}_o$, may be separated into two parts, thus, $_{oo}$ $_o$. What a mercy, then, that not everything in the world is nailed down. ... What a pity that Mill did not also illustrate the physical facts underlying the numbers 0 and 1!
>
> "This proposition being granted", Mill goes on, "we term all such parcels Threes." From this we can see that it is really incorrect to speak of three strokes when the clock strikes three, or to call sweet, sour and bitter three sensations of taste; and equally unwarrantable is the expression "three methods of solving an equation." For none of these is a parcel which ever impresses the senses thus.

Parts of this criticism are mistaken and, indeed, unfair to Mill. Mill refers to many other physical facts beside the one concerning three objects. For example, he says "Ten must mean ten bodies, or ten sounds, or ten beatings of the pulse."[10] Thus, Frege's claim that on Mill's view it would be incorrect to speak of three sensations is mistaken. Indeed, Mill claims that all observable things are characterized by quantity. He says "propositions ... concerning numbers ... are propositions concerning all things whatever."[11] Again, the criticism that Mill's view depends on the fact that physical objects are movable is mistaken also. A collection of objects which impresses the senses thus $_o^o{}_o$ can also be perceived as two adjacent collections $\overset{\scriptstyle{o}}{o}o$. It is not necessary to spatially rearrange the objects.

I agree with Frege however that true arithmetical propositions do not express observable physical facts—that is that in such propositions there is no reference to physical objects or observed objects. This implication of Mill's view is most unlikely given the endless extent of natural numbers and also of the rational numbers. Whether or not there are observed facts concerning collections as large as the number Frege mentions, there surely are true numerical propositions concerning collections larger than any collection that will ever be observed. There are probably true numerical propositions concerning collections larger than any collection of physical objects (assuming that the number of physical objects in the universe is finite). There are true numerical propositions involving rational numbers which are so small that they do not correspond to any physical or observable facts.

A defender of Mill might suggest that while it is not the case that every numerical proposition expresses an inductive generalization, that nonetheless all the numerical propositions of arithmetic can be derived from a small number of axioms—in particular, from Peano's postulates (see section 16). Mill might have alleged only that these postulates are inductive generalizations and that the other numerical propositions are simply logical consequences of these postulates. But, this defense is not very promising. For we may ask, along with Frege, what observed fact does the postulate that zero is not the successor of a number express? Or, we may note that the usual recursive defintions of addition and multiplication presuppose the existence of a particular recursive function. In asserting the existence of such a recursive function, what observed fact are we expressing?

Mill's error, we believe, stems from not distinguishing inductive generalizations from theoretical hypotheses which are empirically supported. Mill assumed that all principles which are empirically supported must be about observable phenomena and so that all such principles are empirical generalizations. However, it may be that the arithmetical propositions which Mill considered are empirically supported theoretical principles. They could be about numbers. However, the fact that such propositions were about numbers would not imply that such propositions are not supported by sensory evidence. Propositions about non-observable entities can be confirmed in the hypothetico-deductive method.

There is another argument of Frege's which appears worthy of comment. It is expressed as follows:

Induction (then, properly understood) must base itself on the theory of probability, since it can never render a proposition more than probable. But how probability theory could possibly be developed without presupposing arithmetical laws is beyond comprehension.[12]

Here Frege is arguing that since the conclusion of an inductive inference is only rendered probable relative to the evidence cited, that inductive reasoning presupposes knowledge of mathematical probability theory. Thus, Frege is contending, that Mill's view that arithmetical principles are inductively confirmed couldn't be correct as it involves a vicious circularity, namely, in order to know arithmetical principles we must know that inductive reasoning is valid. But, in order to know that inductive reasoning is valid we must know that the mathematical theory of probability is true and in order to know this we must know arithmetical principles. This criticism, if it were correct, would apply both to Mill's view as well as to the alternative view which we have suggested. The use of the hypothetico-deductive method of confirmation also shows that the theoretical principles are "probable" relative to the evidence.

In reply to this objection we claim that in the use of the hypothetico-deductive method to show that certain principles are "probable" we are not, in general, using the mathematical theory of probability. The claim that a theoretical principle has been confirmed need not involve the assignment of numerical probability values. Qualitative probability judgements are all that is required. Indeed, due to many other factors it does not look as if the mathematical theory of probability enters into the logic of confirmation anyway. A far more plausible analysis of the logic of confirmation involves what has been called the "cost-benefit dominance" principle. According to this principle, in deciding on the confirmation of scientific principles we evaluate the principles with respect to a range of criteria, e.g., accuracy of predictions, simplicity, coherence with other accepted principles, etc. A theory may be judged acceptable or not acceptable with respect to any of these criteria. Comparisons can be made between theories in that one theory can be judged more or less acceptable than another with respect to these criteria. If one theory is acceptable by reference to all the criteria or if it is more acceptable than any alternative with respect to all of the criteria then we may judge that the theory is confirmed. Numerical judgements are not involved.[13]

Moreover, even where there are numerical principles used in the confirmation of scientific principles, as is the case when statistical methods are employed, there is no vicious circle. The justification of the use of statistical methods may be analyzed in terms of the cost-benefit-dominance approach.

46. I now wish to consider an alternative version of the view that mathematical truths rest on observational evidence. This view holds that certain mathematical terms refer to empirical objects of such a sort that through direct observations of such objects we can obtain certain knowledge. In particular, Hilbert suggested that in number theory we have certain knowledge concerning certain sets of objects which he referred to as "numerical symbols." He represented these objects as follows: 1, 11, 111, 11111. Speaking of these objects as "symbols" is misleading since, according to Hilbert, such objects "have no significance in themselves". Statements of arithmetic such as "3 + 2 = 5" are assertions about these symbols.[14] We can know with certainty that this statement is true because we can observe that the meaningful symbol "3 + 2" designates the same meaningless symbol, namely the symbol "11111", as is designated by the meaningful symbol '5'.

Actually, in classifying Hilbert's view as empiricist, I may be making an error. Hilbert thought that our knowledge of elementary arithmetic was certain. He implies that such knowledge is certain because to obtain such knowledge we use "concrete material finitary methods."[15] I have taken this expression to imply that he regards the symbols '11', '111', etc., as physical objects of which we have knowledge through observation. These methods are finitary in the sense that the number of strokes in such a symbol is finite and so can be completely surveyed. However, Hilbert also speaks of our knowledge of such objects as "intuitive." The use of this term suggests that our mode of knowledge of elementary arithmetic is not through the use of the senses but involves the use of some faculty of intuition.[16] But then we have to ask for a description of this faculty. It may well be that what Hilbert had in mind with regard to our basic arithmetical knowledge is some form of introspection of objects that exist in our mind. If this is the case then his view is clearly similar to that of the intuitionists whom we have considered above. Earlier formalists apparently fall more clearly into the empiricist camp. Frege, in his criticisms of E. Heine and J. Thomas, alleged that according to these thinkers the objects referred to in meaningful numerical statements are visible, tangible, figures.[17] In criticizing Hilbert's views we shall keep this other interpretation in mind.

As noted above, Hilbert thought that our knowledge of

elementary arithmetic is certain because the objects referred to in the meaningful arithmetical statements are "finitary." However, Hilbert recognized that a good deal of what is called mathematical knowledge cannot be reduced to statements about finitary objects. There are statements which appear to be about infinitely large sets of objects, for example, much of Cantor's theory regarding infinite collections. In such mathematical theories there are other signs besides those which refer to finitary objects. Hilbert's view is that such signs are not meaningful. They do not refer to any objects. Expressions containing such signs are not statements but merely "formal structures." He also called such expressions "ideal statements."

At this point two questions naturally arise. The first is, why should such expressions occur in mathematical theories? What role do they play? The second question is, does the occurrence of ideal statements in mathematical theories undermine our mathematical knowledge? That is, in light of the fact that ideal statements play a role in mathematics, can we still justifiably claim that our mathematical knowledge is certain?

Hilbert's answer to the first of these questions is that the ideal statements facilitate logical reasoning. If, in reasoning, one could not infer ideal statements, then the principles of reasoning would be rendered much more complex. Certain argument forms, such as those resting on the law of excluded middle, would not be valid.[18] Ideal statements also serve to simplify communication since the finitary statements which would have to be used in the absence of ideal statements would be cumbersomely long. Consider the statement that there is a prime number between one billion and one billion factorial. In light of the use of the existential quantifier, this is not a purely finitary statement. However, it is acceptable as a meaningful statement because in this case the statement is equivalent to an enormously long disjunction, namely either a billion and one is prime or a billion and two is prime or ... a billion factorial is prime. Further, the proof of this statement, due to Euclid, will obviously apply for any number no matter how large. But, if we cannot use ideal statements in mathematics then we could not say that for any number there is a larger prime number. This latter statement is not equivalent to a finitary disjunction. Since such existential statements are not finitary, if we could not use ideal statements in mathematics then we could not freely apply the principle of excluded middle in proofs. Applications of this principle often involve use of such "ideal" existential statements.

But, is the use of ideal statements in mathematical theory justified? Surely we must be concerned that the deductions which we allow are not such as would permit the derivation of false statements from true premises. Hilbert was concerned about the question of justification. He held that the use of ideal statements is justified providing that we know that the mathematical system which we create by adding the ideal statements to the finitary statements is consistent. Thus, Hilbert required as a condition of acceptability of a formal structure that there be a proof of consistency for the structure.

We may ask however whether a proof of consistency is sufficient to legitimate the use of ideal statements. It seems that it is not sufficient since even if the total set of statements of the mathematical system is consistent it may be that one could derive false conclusions with their use. Surely one wants to know that one cannot derive false conclusions from true premises with the aid of the mathematical system and surely internal consistency is not sufficient to guarantee this result. However, while Hilbert often spoke as if a proof of consistency was all that was needed to justify the use of ideal statements, it is apparent that he had a stronger condition in mind then mere self-consistency. He said "the extension of a domain by the addition of ideal elements is legitimate only if the extension does not cause contradictions to appear in the "finitary domain."[19] In other words, in order to justify the use of ideal statements we must know that the addition of such statements leads to a system which is not merely self-consistent but which does not yield an inconsistency in the subset of finitary statements. In particular, he says, "The problem of consistency ... reduces obviously to proving that from our axioms and according to the rules we set down we cannot get '1 ≠ 1' as the last formula of a proof."[20] Recalling that only the finitary statements are, according to Hilbert, meaningful, we note that what Hilbert's condition amounts to is that, in so far as a mathematical system permits the derivation (proof) of statements as opposed to mere formulas, we must know that we cannot derive any statement which contradicts finitary statements that we know are true. This is quite a strong condition indeed.[21]

Further, the proof that the system does not contradict any known (finitary) statements must not assume any principles except those which are themselves finitary. This notion of finitariness is not of course precisely explained. We may understand it by saying that for Hilbert the proof of consistency must not assume any principle which would not be acceptable to an intuitionist such as Brouwer or Heyting. Alternatively, we could (roughly) explain this notion by saying that in a finitary proof each step is such that either it is known to be true directly through "intuition" or sensory observation of the strokes or which could be established through a perfectly mechanical routine such as through counting.

Mathematical knowledge, on Hilbert's view, consists of finitary statements. The use of ideal statements is justifiable as a means of simplifying proofs and abbreviating statements providing that a proof of consistency of the appropriate sort can be given. In this way Hilbert thought that one could justify mathematical structures which would include, for example, the work of Cantor in set theory and that of Weierstrass in the foundations of analysis, that is, mathematical structures such as real number theory.

Unfortunately a theorem established by Gödel showed that Hilbert's goal is not attainable. Finitary consistency proofs for mathematical theory which is as powerful as real number theory cannot be given. They cannot be given even for much less powerful mathematical principles.[22] Other sorts of consistency proofs have been given for much of mathematics. Whether such consistency proofs have any philosophical value is a matter of debate.[23]

In spite of the failure to attain Hilbert's goal, some philosophers have continued to espouse the view that mathematical knowledge is knowledge of statements about strokes or at any rate of meaningless symbols of some sort. Most notable among these philosophers are Haskell Curry and Abraham Robinson. Subsequently we shall consider their theories to determine whether they yield a satisfactory theory of mathematical knowledge. Prior to such consideration I wish to criticize Hilbert's theory of mathematical knowledge.

47.   Two objections to Hilbert's epistemology immediately spring to mind. He held that mathematical knowledge is absolutely certain and exact. But if, as we have interpreted his view, mathematical knowledge rests on sensory observation of physical objects then it can never be perfectly certain or absolutely exact. Indeed, Stephan Körner has objected to formalism on the grounds that, in his terms, the concepts of mathematics are "exact" whereas empirical knowledge is necessarily inexact.[24] When a concept is inexact according to Körner, then the concept allows for

the existence of cases to which its application is indeterminate, that is to say, the range of the concept is indefinite. For example, we might allow that '11' is an instance of the concept of the number two. But suppose we consider other similar sets of strokes in which the strokes are closer together. Then we shall come to cases in which the strokes are so close together that it is not clear whether we have an instance of the concept of the number one or of the number two.

However, it is not clear that Körner's criticism is correct. We must ask whether the concepts of mathematics are necessarily exact? A concept is perfectly exact if the rules for application of the concept permit of no indeterminate cases—cases to which it is equally in accord with the rules either to apply the concept to the object or to withhold the concept. But, with respect to the mathematical concept of set we may ask, what are the rules for application of the concept? In discussions of set theory the notion of set is often taken for granted and the rules regarding the application of this concept to objects are not explicitly stated. In axiomatic presentations of set theory the basic notion of set is left undefined. I do not see how we can conclude that this concept is exact.

Of course, we can agree that if Hilbert regarded mathematical concepts as perfectly exact and if he regarded mathematical knowledge as resting on sensory observations of strokes then his view is open to Körner's criticism. Empirical concepts cannot be perfectly exact. However, Hilbert could avoid this criticism by claiming only that mathematical concepts are highly exact or that they are exact up to the limits of precision of suitable measuring procedures. Similarly, Hilbert could perhaps avoid the objection that mathematical knowledge cannot be certain if it rests on empirical evidence by claiming that due to the great simplicity of the observations involved our mathematical knowledge can be highly certain.

Alternatively, we might call attention to the point noted earlier that it is not clear that Hilbert thought that mathematical knowledge rested on evidence of the senses. We noted that he speaks of mathematical knowledge as being "intuitive." However, in this case Hilbert's epistemology would be open to the same objection as is directed against intuitionist philosophers. We do not agree that we have a faculty of intuition whose use provides an absolutely unchallengable basis for mathematical knowledge. If, in order to know for example that three is greater than two, we must inspect objects in our own mind, then we do not agree that such knowledge is absolutely certain since the process of inspection is not necessarily infallible. In taking this position we might note that other philosophers who have followed much of Hilbert's way of thinking about mathematical knowledge would agree on this point. Haskell Curry, for example, considered intuition an unacceptable basis for mathematical knowledge.[25] Surely the formalist must hold either that mathematical knowledge is either not perfectly exact and certain or that it rests on some basis other than that provided by either the senses or intuition.

Even if the formalist were to allow that mathematical knowledge is inexact and not perfectly certain, it appears to me to be open to further serious criticism. Let me reiterate that according to Hilbert the subject matter of number theory is certain "symbols." He said "The subject matter of mathematics is ..., the concrete symbols themselves whose structure is immediately clear and recognizable."[26] Earlier formalists also held that the subject matter of mathematics consists in tangible signs. But we may ask, if as the formalists contend the subject matter of mathematics is certain physical objects then we want to know which ones. Is it the ones printed on page 143 of my copy of Benacerraf and Putnam and which look like this: 1,11,111,11111? Or is it the similar ones on the opening page of this section? Or is it the ones on page 143 of some other edition of Benacerraf and Putnam? Perhaps the United Nations Organization should keep a set of such objects stored in a vault somewhere and we could understand numerical symbols such as '2', '3', etc., as referring to those concrete objects. Clearly, all of this is fanciful. Numerals do not refer to any concrete particulars. As we argued in the introduction to this work, numbers are universals.

If numbers are not concrete particulars then the formalist view concerning our arithmetical knowledge must be mistaken also. That is, if the numbers are not concrete particulars then we do not gain knowledge of arithmetical truths through sensory observation of the numbers themselves. Granted that elementary truths could be learned through observation of signs such as those conceived by Hilbert, such truths are frequently not learned in this way. They are learned through observation of other objects such as fingers.

48.   Some formalists have tried to avoid the last criticism by claiming that it doesn't matter what objects we construe mathematical propositions as being about or what objects we

observe. Mathematical truths are determined, they suggest, by the rules for operating on signs—no matter what objects are taken as signs. This view is suggested by formalists in the nineteenth century. For example, Thomas wrote

> The formal conception of arithmetic ... does not ask what numbers are ..., but rather what is demanded of them in arithmetic. For the formalist, arithmetic is a game with signs, which are called empty. That means they have no other content (in the calculating game) than they are assigned by their behavior with respect to certain rules of combination (rules of the game).[27]

More recently this view has been expressed by Haskell Curry.

In Curry's view, "the central concept in mathematics is that of a formal system. Such a system is defined by a set of conventions, which I shall call its *primitive frame*, specifying the following: first, what the objects of the theory ... shall be; second, how certain propositions, which I shall call elementary propositions, may be stated; and third, which of these elementary propositions are true."[28] It doesn't matter what objects are taken as objects of the theory since truth of statements is determined by the rules which determine the primitive frame. Any set of objects taken as the objects of a primitive frame are said to be a representation of the primitive frame. In order for us to think of a formal system we must think of it as represented in some form. However, the formal system is not to be identified with any of its representations. There are characteristics peculiar to the representation of a formal system that are not found in the formal system itself.[29]

Curry gives many examples of formal systems in his work. One that is easy for most philosophers to grasp is a formal system for propositional logic. The objects of the theory, called terms, are for example, p, q, r, s, ... The elementary propositions are usually called well-formed formulas. These include, for example, formulas such as "$\sim p \vee q$", "$p \mathbin{\&} q$", ... The stipulation as to which elementary propositions are true is often done by specifying certain elementary propositions as axioms and in addition indicating certain rules for generating further truths (theorems) from the axioms.

According to Curry, mathematical theories are to be construed as formal systems. This has the advantage, he believes, of eliminating "metaphysical assumptions" from the body of mathematics. By the term "metaphysical assumptions" Curry seems to refer to criteria of truth for mathematical principles which are "vague" and also to assumptions concerning the existence of objects which cannot be verified. Curry criticizes intuitionists for permitting vague criteria of truth to be an essential part of mathematics. The intuitionist must appeal to intuitions and constructions to determine which mathematical propositions are true. But there is conflict amongst intuitions and further how can we tell when we have a veridical intuition rather than merely a strong feeling that some statement is true? Formalism allegedly enables mathematics to avoid this dependence on intuitions. One can, according to the formalist, determine which mathematical propositions are true by appealing to rules for a formal system. Whether a statement is true can be determined, the formalist alleges, objectively by reference to such rules.

Curry also criticizes the platonist for making unverifiable metaphysical assumptions, for example, that there exist infinitely many mathematical objects. On the formalist view, Curry alleges, this assumption can be avoided. Curry also claims that the platonist also lets vague intuitions be essential in determining mathematical truth. For example, Gödel suggests that one must try to settle whether some mathematical axioms are true by reference to intuitions.

But, we may ask, has formalism accomplished its goal of eliminating metaphysical assumptions from mathematics? We note first that formalism has certainly not eliminated metaphysical assumptions from the philosophy of mathematics. As a theory about the nature of mathematics, formalism (Curry's version) is committed to the existence of formal systems and primitive frames. Granted that such objects only exist in concrete manifestations. Still, this only means that Curry's metaphysics is closer to that of Aristotle than to that of Plato. Someone who was concerned to deny the existence of queer entities would not be satisfied with the ontological position outlined by Curry.

Secondly we may ask whether Curry's view does provide an "objective" criterion of truth for mathematical statements. Can we identify the class of true statements in mathematics with the class of axioms and theorems of formal systems? Clearly we cannot make this identification. Any consistent statement can be obtained as a theorem in some formal system. Yet some self-consistent statements are regarded as false. For example, the statement that the function x^2 is *not* a conintuous function of x is self-consistent and so could be a theorem of a consistent formal system. Yet in mathematics texts on calculus we find that the denial of this statement, namely that x^2 is a continuous function of x, is considered as a true statement. Both statements are self-consistent. But both statements cannot be true. Thus, at best, the property of being an axion of theorem of a formal system is not a criterion of truth in mathematics. *A fortiori* it is not an objective criterion of truth.

Thirdly, even if the property of being an axiom or theorem in a formal system did coincide with being true in mathematics, it is not clear that this would be an objective criterion of truth. Whether this would be a criterion of truth that eliminated "vagueness" is dependent on how we can tell whether some statement is a theorem. Is this determined "objectively" by making sensory observations of the representation of a formal system? Then it is subject to whatever vagueness and insecurity accrues to sensory observations. Some philosophers have regarded sensory propositions as more vague and less certainly true than propositions confirmed by "intuition." But, even if we don't wish to take that position, we don't agree that we have any infallible authorities concerning mathematical truth. If we are supposed to obtain knowledge of formal systems by observing representations of such systems then we must allow that our observations may be in error. In observing the representation we may mistake some accidental property of the representation for property which is essential to the formal system itself. At least, Curry's explanation of his theory seems to allow for this type of error. Also, we may simply be mistaken about what the essential features of the formal system are. Indeed, the "metaphysical" aspect of Curry's theory becomes immediately apparent as soon as we ask for a characterization of the essential properties of some formal system. Consider the formal system of logic indicated earlier. What are the essential properties of this system? Presumably the shape of the terms is merely accidental or is it? Would we have the same formal system if we represented the terms by $P,Q,R, \ldots$ instead of by p,q,r? In answer to this sort of question we might be told that the properties of the formal system itself (as opposed to the accidental properties of its representation) are the logical relations between the formulae. Here we see clearly the Aristotelian metaphysics. Curry's view asserts the existence of abstract entities such as formulae and perhaps also of logical relations.

In reply to our contention that his formalist theory does not provide a criterion of mathematical truth, Curry might claim that we have confused mathematical truth with the acceptability of a formal system. References to certain statements of calculus as true reflects, he might say, the mathematician's acceptance of a certain formal system as useful in applications in natural science. In order to develop this answer it is necessary for Curry to indicate how the statements of a formal system are applied in natural sciences (or in some other area). Curry's account of how this is done is quite sketchy. He says that in order to apply a formal system to some subject matter we must first interpret the predicates of the formal system. "In an interpretation we associate them (the predicates) with certain intuitive notions, so that the question arises as to the agreement between the truth of the propositions of the formal system and that of the associated intuitive ones. Acceptability is thus relative to a purpose and a discussion of acceptability is pointless unless the purpose is stated."[30] Curry suggests that classical mathematics, including calculus, is acceptable for purposes in natural science because it is useful in that science. Curry's answer to our objection is then that the formalist does provide a criterion of mathematical truth but it does not provide a criterion of acceptability. He implies that one cannot give an absolute criterion of acceptability since what is acceptable is relative to various purposes. A formal system acceptable relative to the physicist's purposes may not be acceptable relative to other purposes.

But this answer is not satisfactory. Truth is not relative to formal systems. But surely the notion of provability in a formal system is relative. A formula which can be obtained by following the rules of one formal system may not be obtainable in some other formal system. The result of Curry's theory then is not, it might be argued, a criterion of truth but a relativisation of the notion of truth to formal systems. We are tempted to draw an analogy with ethical thought at this point. Often when people are discouraged with efforts to find an absolute foundation for ethical principles they turn to relativism. This does not yield an objective criterion of moral value. Rather it replaces the absolute notion of rightness or goodness with a relative notion, namely rightness in society S. The relativist theory doesn't provide an answer to the question of how we know which ethical principles are true. Instead the relativist implies that no ethical principles are true. Curry should say the same thing with respect to mathematics.

Further, we must note that when the formalist such as Curry, speaks of a proof of a proposition in a formal system, he is not using the term "proof" in the sense of the term which we have discussed earlier in this work (section 32). According to the formalist, a formula is proved if it can be obtained in accordance with the stipulations as to which formulae are axioms. In formal systems in which certain formulae are called axioms, a proof of a formula is a sequence of formulae starting with some axioms and in which each new formula in the sequence is obtained from earlier formulae by application of the rules of derivation and such that the formula to be proved occurs in the sequence. Proof, as so conceived, has no epistemic value, i.e., it does not establish that a formula corresponds to reality. The theory of proof and of truth to which formalism leads then is quite different from the ideas on these matters which were part of Hilbert's thought.

The view that mathematical statements are merely formulae in formal systems is not an accurate reflection of the way that most working mathematicians think about the statements of mathematics. Consider the fact that mathematicians in their work often provide diagrams to represent the content of the statements that they make. Sets are represented as closed regions on a plane surface. Geometrical figures are used to represent triangles, etc. If the statements were considered merely as meaningless objects in a formal system then how could any diagram be appropriate as a representation of the content of the statement. If the formalist view of the nature of mathematics is correct then many mathematicians must be said to have very misguided ideas as to the nature of what they are doing. But this suggestion is not very plausible.

49.   A philosophical position close to those of Hilbert and Curry has also been expressed by Abraham Robinson. Robinson affirms that statements in mathematics which would normally be interpreted as affirming the existence of infinite totalities should be regarded as ideal or "not literally meaningful." Robinson does not go so far as Hilbert in holding that finitary mathematical statements are about physical objects. Robinson considered views such as Hilbert's to be nominalistic and he rejected the nominalist point of view. He said "I do not feel compelled to follow the nominalists who seem to have little trouble in grasping the notion of an individual but feel incapable of proceeding to the notion of a class."[31] It appears then that Robinson is willing to allow the existence of such queer entities as classes. However, he is not willing to allow that there are infinitely large classes all of whose members actually exist. His view is like Hilbert's in that he holds that such statements are "ideal" or like Curry's in that he allows that such statements are uninterpreted parts of a primitive frame.

He holds that even though such statements are not literally meaningful we are justified in retaining them as part of our mathematical theories due to the fact that such theories prove to be extremely useful when they are applied in natural sciences. Here again he wants to follow the view of Curry in holding that a mathematical theory can be acceptable even if it is not literally meaningful. He held that "A mathematical theory is acceptable if it can serve as a foundation for the natural sciences."[32]

We have criticized Hilbert's version of formalism by arguing that it is a mistake to regard mathematical terms as referring to physical particulars. Curry's version of formalism avoids this objection but, as we argued above, it provides no satisfactory theory of mathematical truth or knowledge. Curry's view, as we noted, yields the consequence that mathematical truth is relative. I believe that there is a fair interpretation of Robinson's view which avoids both of these objections.

In order to present this interpretation let us consider a theory from some branch of a natural science. It might be a theory of gravitation from physics or a theory concerning the flow of blood in a mammal's circulatory system. Let us call the theory T. Associated with T is a mathematical theory M. M is an essential part of T in that logical inferences which are made by scientists using T would be invalid unless principles included in M are presupposed. Let us suppose for the sake of this example that M is the theory of real numbers to which we have referred in several places in this work. Now, let us suppose that T is formulated in a language of first-order predicate logic. In this language there will be certain predicate constants, let us say $P, Q, R,$ and S. Certain relationships involving these predicates will be determined by the axioms of T. Robinson's theory is that many of the statements of T may be literally true. Included among these statements which may be true are even many statements from the mathematical part of T. However, in certain statements of T, the predicate constants would, if interpreted literally, refer to infinite totalities. Robinson's view is that such predicates cannot be interpreted literally. These statements which, if true, would imply that there exists an infinite totality are to be regarded as ideal. Robinson is willinging to allow that a theory as a whole can be acceptable even though some of its statements are ideal. This is equivalent to saying that a theory can be acceptable even though some of the terms occurring within the theoretical statements have no literal interpretation.

If one thinks about the above theory T which includes real number theory M as a part, some questions arise as to how Robinson's view is to be understood. In particular, are we to regard all of the statements of real number theory as ideal since taken as a whole these statements imply the existence of infinite totalities. This would be the simplest course to take. It would imply that such terms as real number, addition, negative, upper bound, etc., are not literally meaningful. However, Robinson has indicated that he regards statements about finite totalities as both literally meaningful and as *not* referring to physical objects. Thus, it would appear that we should regard some of the axioms of real number theory as meaningful. But this poses problems.

Suppose for example that we say that the field axioms are meaningful since they do not imply the existence of an infinite totality but that the axioms asserting the existence of an ordered field are not literally meaningful since these axioms do imply the existence of an infinite totality. But now we find that most of the predicates for the axioms of an ordered field also occur in the axioms for a field. It appears that we must therefore hold that while these predicates are meaningful in the context of the field axioms, they are no longer meaningful (literally meaningful) in the context of the axioms for an ordered field. Now it might be argued that this complexity in Robinson's theory is not, of course, an objection that need be taken seriously. The situation in regard to mathematical terminology it might be said, is entirely analogous to other terms in physical theory. Some terms in the context of physical theory may be literally meaningful but the same terms may also occur in the theories advanced in writings of science fiction. Very often, in these contexts, the same terms do not make sense. Such statements cannot be understood as having the referents that they have in their scientific uses. However, I am not sure that this remark concerning the fact that the same terms occur in both scientific and fictional contexts suffices to answer the difficulty which we have raised in Robinson's theory.

When one thinks about the use of scientific terminology in fictional contexts and asks why such uses sometimes make no sense several answers come to mind. Sometimes the use of such terms in fictional contexts makes no sense because the statements made in such contexts contradict accepted principles of natural science. Other times statements in fictional contexts make no sense because such statements are, as it were, simply thrown into the dialogue. No effort is made to develop a coherent theoretical framework for the statements. In the absence of such a framework the statements have no literal meaning. For reasons such as these, terms and statements in fictional contexts may fail to be literally meaningful even though the same terms and statements are literally meaningful in other contexts. However, these reasons do not apply to the case of theory T. The mathematical principles of real number theory contradict no accepted statements of laws of nature and statements in which reference is made to infinite totalities are integral parts of coherent theoretical frameworks.

We have been discussing a difficulty with Robinson's theory as follows: Scientists accept physical theories such as theory T which include real number theory. According to Robinson's theory we should regard the field axioms by themselves as meaningful and the predicates used within the statements of these axioms as having a literal interpretation. However, we cannot regard the axioms which assert the existence of an ordered field as meaningful and so the terms which occur in these statements have no literal interpretation. The problem is that the same terms occur in both axioms. Thus Robinson's view seems to commit him to saying that certain terms both have and do not have a literal interpretation. We have considered one possible answer to this difficulty. This is to say that the terms can be meaningful in some contexts and meaningless in others because in some contexts the use of terms is analogous to the use of terms in fictional contexts. I have found this answer to the difficulty unsatisfactory because the factors which render the use of terms meaningless in fictional contexts are not present in the context determined by the inclusion of real number theory as part of scientific theories.

There is however another way for Robinson to answer the difficulty in question. He can say that the statement that theories in natural science include real number theory as a part is true but also that it is only part of the truth. He could hold that such theories contain in addition mathematical theories which are finitistic. Such theories would assert the existence of only a finite number of mathematical entities. They would also contain terms which signify mathematical operations such as addition, and multiplication. They would also contain terms such as "lower bound", and "less than", etc. He could hold that since these terms derive their meaning within a finitistic theoretical context then, in spite of any similarity to certain terms of real number theory, they are not the same terms. Thus, he does not have to say that the same terms have a literal meaning and no literal meaning.

Of course, natural scientists when they make use of mathematical theories do not usually make clear whether the terms they are using are to be understood in the finitistic sense or have no literal interpretation because they occur in an essentially infinitistic context. Thus Robinson's answer may seem a bit artificial. Nonetheless, he could say that for many scientific purposes it makes no difference whether the mathematical terms are understood finitistically or otherwise and so natural scientists tend to be careless about this distinction. When it becomes important for philosophical purposes to determine which scientific statements are meaningful, the philosopher of mathematics can make the decision by considering the contexts in which the statements are made.

While Robinson can answer the above difficulty, I believe that there is still reason not to accept his theory. To see this let us note again that Robinson agrees that finitistic mathematical statements assert the existence of queer entities. His view is not only close to those of Hilbert and Curry. It is also close to that of Gödel. With regard to ontology Robinson differs from Gödel only in regard to the existence of infinite totalities. But what is the epistemological basis for our mathematical claims according to Robinson? In the paper cited above he does not answer this question and I am not aware that he has done so in any other place. He does, in the above cited paper, indicate that a mathematical theory is acceptable if it is useful in the natural sciences. But, since real number theory is acceptable by this criterion, he would not identify what is acceptable with what is known. In the absence of a satisfactory theory of knowledge the ontological theory espoused by Robinson (and also by Curry) is arbitrary. We have rejected both Gödel's and Brouwer's theories for epistemological reasons and while we cannot reject Robinson's theory because he has an unsatisfactory epistemology, we are justified if we withhold our acceptance from his view since he has expressed no answers to the basic question of how mathematical statements are known. Robinson avoids the difficulties of Hilbert's epistemology but he does so by avoiding fundamental questions which Hilbert tried to answer. Since we have been searching for an answer to these questions we shall have to look beyond the views of Robinson. Possibly, within the framework provided by a suitable epistemological theory, Robinson's ontological view would be acceptable.

50. Conclusion: In this chapter we have been considering empiricist theories of mathematical knowledge. Mill's theory is mistaken in regarding mathematical principles as empirical

generalizations. But the criticisms of empiricism developed by the logical positivists and by Frege are mistaken also. Mathematical principles may be refuted by empirical evidence. We have also considered the views of formalists in this chapter. Hilbert's views are mistaken since mathematical objects are not physical particulars. Later formalists such as Curry and Robinson avoid this objection but, in the end, they provide no satisfactory account of mathematical knowledge. Curry does not explain how knowledge of primitive frames is obtained and his view implies that mathematical knowledge is relative to formal systems. Robinson, as I understand him, avoids this objection. But, in the end all Robinson tells us is that we have no knowledge of infinite totalities. He does not provide an account of other parts of mathematical knowledge.

NOTES

1. Mackie, (50).
2. Quine's views are expressed in (70) and (69). Putnam's views are expressed in (68).
3. Mill, (55) p. 165–6.
4. Hempel, (27) p. 367.
5. Ayer, (1) p. 75–6.
6. Hempel, (28).
7. Mill, (55) p. 163.
8. Arguments against the positivists along the line I have here suggested have been proposed by Stephan H. Levy in (46).
9. Frege, (19) pp. 9–10.
10. Mill, (55) p. 163.
11. *Ibid.*, p. 163.
12. Frege, (19) pp. 16–17.
13. For discussion of the cost-benefit principle see Michalos, (54).
14. Hilbert, (33) p. 143.
15. *Ibid.*, pp.142–3.
16. This interpretation of Hilbert is favored by Kitcher in (36).
17. Geach, (20) p. 182 f.
18. Hilbert, (33) p. 145 f.
19. *Ibid.*, p. 149.
20. *Ibid.*, p. 149.
21. This point is noted by Kitcher, (36) p. 102.
22. I refer of course to Gödel's second theorem in which he established that if a mathematical system such as Hilbert had in mind is consistent, then a proof of the consistency of the system within the system cannot be given. Any proof of consistency would have to be in a stronger system which, in turn, would need to be justified by a consistency proof.
23. See Resnik, (77) pp. 133–147.
24. Körner, (38) p. 159 f.
25. Curry, (17) p. 153.
26. Hilbert, (33) p. 142.

27. This quotation from Thomas is found in "Frege against the Formalists" in Geach, (20) p. 183.

28. Curry, (17) p. 153.

29. Curry, (16) p. 30.

30. Curry, (17) p. 155.

31. Robinson, (78) p. 230.

32. *Ibid.*, p. 234.

Chapter Eight

An Empiricist Theory of Mathematical Knowledge

51.   In section 31 we considered briefly a skeptical view of mathematics, namely the view of Hans Vaihinger. We dismissed that view on the grounds that some people have mathematical knowledge but that if Vaihinger's view were correct then nobody would have mathematical knowledge. While I believe that the position taken contrary to Vaihinger is correct, it is worthwhile at this point to again consider the possibility of adopting a skeptical position. What makes it appropriate at this point in our work is that, as a result of our consideration and rejection of the epistemological views of Gödel, Brouwer, Hilbert and Mill, skepticism may now appear more plausible. Furthermore the development and acceptance of the validity of non-Euclidean geometries has induced many people to adopt skepticism as a philosophy of mathematics. Since it is appropriate to consider skepticism at this point in our work we shall consider arguments based on these geometrical developments also.

Let us, for the moment, take skepticism to be the view that knowledge is impossible. Skepticism with respect to mathematics then is the view that it is impossible for anyone to know any mathematical principles. Now, it might be argued that since the theories of knowledge considered above are all false that it is impossible to have knowledge of mathematical axioms and, that if it is impossible to have knowledge of mathematical axioms then it is impossible to have any mathematical knowledge. Thus, someone might argue that skepticism with respect to mathematics is true. Strictly speaking of course, this argument is unsound. The conclusion that mathematical knowledge is impossible would follow from the falsity of the above theories of knowledge only if the above theories exhausted all possible theories of mathematical knowledge. However, the above theories are not exhaustive with respect to theories of mathematical knowledge. At the very least there is one more empiricist theory of knowledge which is worthy of serious consideration. Further, it might be argued that we have not established even that the above theories of knowledge are all false. Granted objections have been raised and discussed with respect to each of these theories but in each case a defender of the theory may believe that the objections do not show his theory to be false. He may construe the objections only as difficulties which he must overcome in developing his theory. Alternatively he may deny that the objections have any validity by questioning the premises which we have adopted. Nonetheless, as noted above, the difficulties with the above theories of knowledge may incline some people to accept a skeptical position. This has indeed been the case in a recent work in the philosophy of mathematics by Stephan Körner.

Körner does not argue that since the above theories of knowledge are false then skepticism must be true. Indeed he doesn't even use the term *skepticism*. However, after considering and criticizing a number of theories of knowledge, including intuitionism and formalism, Körner adopted a theory about the nature of mathematics which implies that skepticism is true. Körner's theory concerning the nature of pure mathematics is expressed in the following quotations:

> If we compare "There exists a piece of copper" and "There exists an immortal soul", on the one hand, with "There exists a Euclidean point" on the other, we see that the grounds for these existence statements are quite different. Consistency of the constituent concepts is necessary in all cases. But whereas we can make objects for "Euclidean point" available by decision or postulation, we cannot do this for "piece of copper" or for "immortal soul."
>
> The freedom to postulate the availability of Euclidean points implies the freedom to postulate their non-existence. This means that although the statements "There exist Euclidean points" and "There do not exist Euclidean points" are incompatible, this incompatibility does not imply that at least one of them is false.
>
> The same remarks apply to mathematical existence-propositions in general ... the pure mathematician can make the objects of his self-consistent concepts available by his own *fiat*.[1]

But, to say that the mathematician can create the objects of his concepts by his own decision or fiat is in effect to say that the objects referred to in pure mathematics are imaginary. There is no way to create any objects by fiat, unless of course one is speaking of creation in one's imagination. Körner's view then is that the objects referred to in mathematics are not real. They are, to use Vaihinger's term, fictions. This is no more than to say that mathematical existence propositions are false. And, if these propositions are false then they certainly cannot be known.

(Following well-established philosophical tradition, I hold that if a proposition is known then it is true.)

It might be argued that Körner's skepticism is limited to mathematical-existence propositions, but that he regards other mathematical propositions as true and so as capable of being known. Indeed, if he were thinking of existence propositions as solely propositions of the form "There exists an object such that. ..." then it would appear that he would have to hold that some mathematical propositions are true, namely the denials of these existence-propositions. However, as he does not spell out the category of existence propositions we cannot assume that he intended it to be limited in this way. Indeed this would be an absurd view as it would imply that all purely universal propositions are true. For example, since "There exists a non-isosceles Euclidean triangle" would be false, it would be true that all Euclidean triangles are isosceles.

Further, Körner's remarks concerning the incompatible propositions "There exists Euclidean points" and "There do not exist Euclidean points" suggests that he did not really intend to say that either of these propositions is false. If neither of these propositions is false, then neither of their denials is true. This suggests that Körner's view is that mathematical propositions have no truth value. This view also implies skepticism with respect to mathematics. For, if mathematical propositions have no truth value then *a fortiori* they are not true and consequently they cannot be known.

To ask us to refute skepticism in the sense of showing that it is false is to ask for too much. However, to ask us to refute skepticism in the sense of showing that it is ungrounded is a reasonable request. We have already argued that refutation of the theories of knowledge we have considered in earlier chapters does not provide sufficient grounds for skepticism. Let us next turn to consideration of the acceptance of non-Euclidean geometries to determine whether his provides a sufficient basis for adopting a skeptical philosophy of mathematics. In light of the example he used in the above quotation it is apparent that these developments in geometry have influenced Körner's thought on these matters. In our discussion we shall also make reference to the existence of alternative set theories. As in the case of geometry, the development of alternative set theories has led some people to take a skeptical position.

52. The problem posed by the acceptance of non-Euclidean geometries as alternatives to Euclidean Geometry arises because of the claim that the various geometries are incompatible with each other and also that mathematically there are no reasons for saying one of these geometries is true and that the others are false. Clearly, assuming that these geometries are incompatible, we cannot say that they are all true. Since there is no mathematical reason for saying that one is true rather than the others, it is tempting to conclude that none of them are true. Once this position is taken it is natural to go further and say that what the student of geometry does is to deduce the logical consequences of various sets of hypotheses. Whereas Euclidean geometry normally includes in its postulate set the claim that there exists only one line parallel to a given line through a given point not on the line (Playfair's postulate), Riemanian geometry adopts the postulate that there are no lines parallel to a given line through a point not on the line. Each geometry consists of its distinct set of postulates and the theorems deducible from them. Having made this claim with respect to geometry then it is a natural step to generalize it for all mathematical theories. We seem inevitably to be led to a postulationist point of view after all.

However, it seems to me that to accept the postulationist or skeptcial point of view would be to jump to an unwarranted conclusion. There is an alternative position which, I believe, is warranted by the evidence. There are two things to keep in mind with respect to this position. First, I claim that in light of the fact that it is so useful in scientific applications, we are warranted at this time in accepting real number theory as true. Secondly, geometry is reducible to real number theory by identification of geometrical points in a plane with pairs of real numbers, etc., as is normally done in analytic geometry. The axioms of Euclidean geometry can be regarded as *true of* certain structures whose existence is implied by the axioms of real number theory plus a certain amount of set theory. The axioms of Riemanian geometry are *true of* alternative structures, and similarly for other non-Euclidean geometries. The student of Euclidean geometry studies the nature of structures which satisfy the axioms of real number theory but which also satisfy certain special postulates. The same applies to the student of other non-Euclidean geometries. On our view all of the geometries in question, both Euclidean and non-Euclidean, are true, albeit they are *not* incompatible with each other due to restrictions in the scope of their principles. The acceptance of the theory of relativity in modern physics has not led to the conclusion that Euclidean Geometry is false. It has shown only that certain non-Euclidean structures provide a better model of spatial relations.

The existence of alternative axiomatic set theories poses a similar problem for one who claims that there is mathematical knowledge. The alternative set theories are mutually incompatible so they cannot all be true. Mathematically, there is no reason for saying that one is true and that the others are false. Here, we cannot solve the problem in the same way we did in the case of geometry. For, while the geometries are each less general in scope than is real number theory, real number theory is certainly less general than some of the set theories. For example, the principles of classical real number theory can be derived within Zermelo–Frankel set theory or within Russell's set theory including the axiom of reducibility.

In this case I would suggest that none of the alernative set theories are true. Perhaps, as Gödel has suggested, a new axiomatic set theory will be developed which is clearly more acceptable than any of the axiomatic set theories currently in existence. My reasons for saying this are as follows: Mathematicians have found various set theoretical principles very fruitful in many branches of mathematics. However these principles are not formulated within any axiomatic theory. The principles have not been stated with the type of precision which is required for axiomatic formulation. Mathematicians refer to these principles as naïve set theory. Mathematicians regard the principles of naïve set theory as true. The various axiomatic set theories can be looked upon as efforts to formulate the principles of naïve set theory in a consistent axiomatic form. However, none of the axiomatic set theories is completely satisfactory in this regard. Some of the axiomatic theories are inconsistent with principles accepted as part of naïve set theory or real number theory. For example the set theory developed by Quine and called *New Foundations* has been shown to be inconsistent with the axiom of choice. In the Zermelo–Frankel set theory there is no universal set. Russell's theory of types is not incompatible with real number theory but it contains the unintuitive axiom of reducibility. In the case of the axiomatic set theories, we have a number of alternative set theories none of which is completely acceptable. But this gives us no reason for skepticism with respect to the principles of real number theory or naïve set theory.

In discussing these difficulties with respect to the view that we have some mathematical knowledge, I have assumed that this view is correct. It is now time to try to justify this assumption. I will do this by briefly developing the empiricist theory of mathematical knowledge which I have hinted at above.

53. Let us start explaining our theory of knowledge by distinguishing between principles which are known because they have been proved and principles which are known without proof. We shall refer to knowledge of the former sort as derived and knowledge of the latter sort as basic. Clearly, there is no intrinsic difference between basic knowledge and derived knowledge. Principles which have been considered as basic may subsequently be proved and will then come to be considered as derived knowledge. Also principles which have been considered as derived may later, in consequence of a systematic reorganization of mathematical knowledge come to be regarded as basic. The distinction between basic and derived knowledge is clearly relative to a systematic organization of the knowledge. However, this is not to say that the organization is a purely arbitrary matter. Clearly, some principles are more general than others and, in light of the logical strength which accrues to them as a result of their generality, may be more appropriate as basic principles than others. However, we cannot say that the basic principles should always have the greatest logical strength of which we can conceive. Clearly, our basic principles must not imply everything. Further, principles whose implications are too widely divorced from applications in empirical sciences can hardly be justified by empirical evidence. Such principles are properly regarded as being of interest from a purely hypothetical point of view.

As indicated in the last two sentences of the last paragraph we regard our knowledge of basic principles as resting on the widespread application of such principles in theories from the natural or social sciences which are in turn confirmed via sensory observations. In other words, in our view basic mathematical principles are known via an inferential process which accords well with the hypothetic-deductive pattern of reasoning which has been widely discussed in recent years. The basic pattern of reasoning here is as follows: From a set of hypotheses and assumptions about particular experimental or observational set-ups plus theoretical assumptions further observational consequences are deduced by valid forms of argument. Given that certain further conditions are satisfied, verification of the observational consequences through sensory observations confirms to some degree the theoretical assumptions which enter into the deduction. I wish briefly to discuss the further conditions which must be satisfied and to say something concerning the notion that the theoretical assumptions are confirmed to some degree.

Clearly, the verification of the observational consequences deduced from the theoretical assumptions and other hypotheses confirms the theoretical assumptions only if the theoretical assumptions play an essential role in the deduction. If the observational consequences could be deduced from the other hypotheses alone then the theoretical assumptions would receive no confirmation even though the observational consequences have been verified. Further, the degree of confirmation which some theoretical assumptions receive from the verification of some of their observational consequences is *not* independent of alternative theoretical assumptions which may provide rival explanations of the same observational consequences. Where the same observational consequences may be explained in several distinct and incompatible ways then the theoretical assumptions entering into any such explanation receive little or no increment in confirmation from the verification of those consequences. On the contrary, where some observational consequences can be accounted for only on one set of theoretical assumptions then the verification of those consequences is strongly confirming for those theoretical assumptions.

I do not believe that one can formulate very precise rules spelling out in detail procedures leading to theoretical assumptions being confirmed to high degree. For one thing, it appears that degree of confirmation cannot be assigned any numerical measure. To hold that degree of confirmation could be assigned a numerical measure and that our basic knowledge rested on empirical confirmation would be to open ourselves up to the objection that Frege directed against Mill, namely that in order to justify any propositions at all in an empirical way we would have to presuppose mathematical knowledge and so it would be circular to then claim that mathematical knowledge is confirmed in an empirical way. Further, Putnam has argued that "the actual inductive procedure of science has features which are incompatible with being represented by a measure function."[2] He argues that an inductive procedure, to be satisfactory must be capable of showing that a hypotheses H is true if, in fact, H is true, but that if the inductive procedure is in accord with some numerical measure function then there are hypotheses which will never be confirmed even if they are true.

To see how mathematical assumptions function in the derivation of observational consequences from theoretical assumptions let us consider a simple example. Our theory in this case relates to an association between distance, average velocity and time. The theory tells us that the distance a moving object travels is equal to the average velocity travelled during a time interval multiplied by the time interval. Let us represent this theory as follows:

$$T: \quad d = vt$$

In apply this theory to a simple problem special additional assumptions may be made. For example, we may assume that the acceleration during the interval of time in question is constant. Thus the average velocity for the time interval will be equal to half of the sum of the initial and final velocities. Let us represent this as:

$$A(1) \quad v = \frac{v_0 + v_f}{2}$$

Suppose now that we are considering the following test of our theory T. Using the theory we should be able to determine the height of a tower from which a stone is dropped providing that we can measure the time it takes for the stone to fall and providing that we can determine the average velocity at which it fell. To determine the average velocity we use special assumption $A(1)$ and assume that its initial velocity is 0 feet per second and that we measure the final velocity as n feet per second. We assume also that we measure the time it takes for the fall and find it to be m seconds. From this observed data plus the special assumptions we determine that the average velocity is $\frac{n}{2}$ feet per second. Using this information plus our theory T we can infer that the distance equals $\frac{n}{2}$ times m feet. Where do mathematical principles play a role in the derivation of this observational consequence? Clearly, such principles play a role in determining the average velocity where the initial and final velocities are added and divided by 2, and also in the determination of the distance d which is determined by multiplying the average velocity and the time. Thus, in addition to the special assumption $A(1)$ there are additional mathematical assumptions which are utilized in the derivation of the observational consequence. Verification of this consequence by an independent determination of the height of the tower confirms the theory T, the special assumption $A(1)$ and the mathematical assumptions as well.

In this example, of course, the mathematical assumptions used involved only the addition, mutliplication and division of rational numbers. If all examples in which theories were tested in physics involved only operations on rational numbers then we could claim perhaps that the theory of rational numbers had received some confirmation but that the theory of real numbers had received no confirmation. In this case the mathematics that we could claim to know on the basis of observation would be closer to that mathematics which, according to Brouwer, we are entitled to claim to know. However, in many physical problems the mathematical assumptions made include not only the assumption of the existence of rational numbers but in addition the assumption of real numbers and of the validity of mathemtical operations on these numbers. For example, consider a theory about the amount of material in the part of the Earth's interior known as the core. The volume of this material is the volume of a sphere and, as is well known, the expression of this volume involves the irrational number π, namely $\frac{4}{3}\pi r^3$. If, in deriving observational consequences from a theory which employs the above principle for the volume of a sphere we operate as we did before by adding, dividing, etc., then we are assuming that these operations are valid for real numbers and if the observational consequences are verified then the theory of real numbers receives some degree of confirmation. Since, we claim that on the basis of empirical confirmation we have knowledge of real number theory, we are claiming to know a logically stronger theory than that which, according to the Intuitionists, we can legitimately claim to know.

54.   An obvious objection that someone might make with respect to the empiricist view which we have just explained is to claim that mathematical knowledge is certain whereas empirical knowledge is uncertain. If this is the case then mathematical knowledge claims cannot be completely justified in accord with the above empiricist theory. In support of this objection it may be pointed out that whereas many scientific theories which utilize mathematical principles as above indicated have been rejected on the basis of observational evidence, the mathematical principles, e.g., of calculus, have not been rejected. If the evidence for mathematics is in accord with the theory that I have indicated above then, it may

be claimed, the disconfirmation of a scientific theory through failure to verify its observational consequences should lead to disconfirmation of the mathematical principles as well. Since this has not happened, it is alleged, we may conclude that the evidence for mathematics is not empiricial.

Our answer to this objection is to claim that mathematical knowledge is not certain. We cannot prove beyond any doubt that the principles of real number theory are consistent. If an inconsistency is discovered then they would have to be modified or even rejected. Further, it is quite conceivable that the principles of real number theory would be disconfirmed in an empirical way. I suggest that this might happen in the following circumstances: Suppose that in some area of physical theory, say the theory regarding motion of projectiles, we found that (1) we were regularly unable to verify the empirical consequences of the theory (consequences derived with the aid of theorems of real number theory), and (2) that we had an alternative physical theory in which the mathematical principles differed from the principles of real number theory in some way and that this theory was confirmed through observations. (We may suppose, for example, that the mathematical theory postulated the existence of objects and operations on those objects which were non-commutative, or non-distributive, thereby contradicting the algebraic properties of the real numbers). This would count as a disconfirmation of real number theory. If this same situation occurred not only in the theory of projectile motion but also in many other areas of science then eventually real number theory would be abandoned.

Our answer to this objection also suggests the explanation of the fact that while real number theory is not known to be true with certainty, our confidence that real number theory will not be abandoned is indeed justified. The progress of science since the development of real number theory in the nineteenth century has seen the abandonment or modification of many theories. The alternative theories which have been adopted have included the principles of real number theory. While the abandonment of real number theory is conceivable, the fact that it continues to be used in conjunction with many theories in the natural sciences, even with theories that are incompatible with each other, strongly suggests that real number theory is true.

In holding that mathematical knowledge is not certain we set ourselves in opposition to an ancient tradition in philosophy—a tradition dating back to Platonic times. Defenders of this tradition might object to our view that the notion of uncertain or fallible knowledge is consistent. They might say that to allow that knowledge is fallible is to allow that it is possible for a person to know a proposition P while not being certain that P is true. If a person is not certain that P is true then, so far as he is concerned, it is possible that P is false. Thus, the concept of fallible knowledge implies that it is possible that a person knows a proposition that P and at the same time that it is possible that P is false. We might represent this in symbols as follows (letting 'M' represent "it is possible that" and '$K_a p$' represent "a knows that p" and "Fp" represent "p is false").

(1) $M(K_a p$ and $MFp)$

But, the objector may claim, this is paradoxical. He might claim that a person who says "I know that p but it is possible that p is false" is contradicting himself. For the objector may argue, for a person to say "it is possible that p is false" implies that he does not know that p. Thus to say "I know that p but it is possible that p is false" implies "I know that p but I do not know that p" and this is clearly self-contradictory.

In reply to this objection to the notion of fallible knowledge, I would urge that the objector is mistaken concerning the implications of "it is possible that p is false". "It is possible that p is false" does not imply that p is false. Nor does it imply that the speaker does not know that p. If a person says "It is possible that p is false," he implies only that he is not certain that p. In saying that we are not certain that p we are not saying that we have any reason to doubt p other than the fact that we have no method of establishing p that we know to be infallible.

Notice that in saying that it is possible for a person both that he knows a proposition p but that it is possible p is not the case, we have not held that it is possible that person knows p but that p is not the case. We affirm that if a person knows a proposition p then p is the case. We allow that it may happen that a mathematician thinks he knows some mathematical proposition p and that subsequently a counter-exmple to p is discovered which shows that p is false. In this case we should say of the mathematician that he did not know that p and that he never knew that p; he merely thought he knew that p. Using the symbols that we introduced above, we affirm the following:

(2) It is not the case that $M(K_a p$ and $Fp)$.

In other words we hold both that

$$L(\text{if } K_a p \text{ then } p)$$

and that

it is not the case that (if $K_a p$ then Lp)

(where 'L' represents "it is necessary that").

The view of mathematical knowledge which we have defended here resembles that discussed earlier of J.S. Mill in regard to Mill's contention that mathematical knowledge rests on empirical evidence. However, we do not agree with Mill's analysis of that evidence. Mill's analysis implied that arithmetical knowledge is simply a generalization of observed instances in which sets of physical objects are counted. But Mill's view is inadequate for two reasons. First, it takes no account of the logical relations between mathematical propositions. Second, there are many mathematical propositions such as principles of calculus which are clearly not generalizations of observed instances. The view of mathematical knowledge which we have sketched above avoids both of these objections.

55. Hilary Putnam has argued that the confirmation of real number theory may not be as great as I have suggested since a reformulation of principles of the natural sciences is possible which involves quantification over the domain of rational numbers but not over the domain of real numbers. If this reformulation could be carried out then we would not be warranted in affirming the existence of real numbers by empirical evidence. According to Putnam

"At first blush, the law of gravitation (we shall pretend this is the only law of physics, in the present essay) requires quantification over real numbers. However, the law is equivalent to the statement that for every rational δ, and all rational m_1, m_2, d, there is a rational $\in$ such that

if $M_a = m_1 + \delta$, $M_b = \pm \delta$, $d = d_1 \pm \delta$ then
$$F = \frac{gm_1 m_2 \pm \in}{d_1}$$

and this statement quantifies only over rational numbers."[3]

However, Putnam's argument that physics could dispense with real number theory in favor of rational number theory is not sufficiently complete to be persuasive. Let us grant, for the moment that the laws of physics (and of other natural and social sciences) could be expressed in the manner Putnam suggests. And let us grant that the formulation of the laws does not require any irrational constants (a difficulty which Putnam notes but does not consider in the above context). We are still not in a position to say that we can dispense with real number theory in favor of rational number theory. The difficulty that may prevent this move is that the principles of real number theory may be assumed in the derivation of observational consequences from scientific theories even if only rational numbers are used in the expression of such (scientific) theories. For example, in the derivation of observational consequences from theories in physics, many principles of differential and integral calculus are assumed. But these are principles which can be proved only on the assumption that set of numbers is a complete ordered field. Since the rational numbers are not complete, the principles of the theory of rational numbers do not provide a justification for these principles of calculus. Indeed, the theorems of calculus will be false if only the rational numbers exist. In light of the fact that scientific inferences presuppose the truth of principles of real number theory we are not persuaded that science could dispense with these principles merely because scientific laws can be reformulated so as not to explicitly assert the existence of real numbers.

56.   We have argued for the existence of queer entities in essentially the following way: to give satisfactory explanations of empirical data physicists must make mathematical assumptions. These assumptions are justified by the same empirical evidence which justifies the non-mathematical assumptions which enter into such explanations. Among the non-mathematical assumptions are such principles as Newton's laws of motion and gravitation, the postulation of genes by geneticists, the postulation of a circulatory system by Harvey, etc. Our argument can be criticized in several ways. It can be claimed that even though the acceptance of mathematical principles is justified by empirical evidence, this does not show that queer entities exist because mathematical principles do not really assert or imply the existence of such entities. We have been at great pains to rebut this criticism in the first four chapters of this work where we tried to show that all efforts to construe mathematical principles or the use of mathematical principles in other sciences as not implying the existence of queer entities were open to serious criticism. Mathematical principles assert that numbers, sets, sequences, etc., exist. Such statements cannot be

satisfactorily interpreted in such a way as to imply that they do not really mean what they say.

But our argument is also open to the following criticism. It may be admitted that in explaining phenomena scientists make mathematical assumptions and even that their making such assumptions is justified by the empirical evidence, i.e., by the verification of the observational consequences of their theories. Nonetheless, it may be claimed that the justification which such verification yields for the use of mathematical assumptions does not warrant believing that the principles are true. In other words the objector is claiming that there is a valid distinction between accepting the principles merely for use in natural science and accepting them in the sense of believing that they are true. He may grant that the verification of observed consequences justifies our using such principles but not our believing that they are true. He may claim that if we conclude that the principles are true on the basis of the empirical evidence we are making a leap of faith. This objection, I believe, gets to the heart of the conflict between the view which we have been espousing (that there are queer entities and that we are justified in claiming this on the basis of empirical evidence) and the view which we have called fictionalism (and have attributed to Vaihinger and Körner). The fictionalist refuses to accept the argument that because mathematical principles are necessary in explaining the data of science the principles are true.

The critic of our argument for the truth of mathematics suggests by his criticism that while empirical verification justifies the use of mathematics it is not sufficient to justify the claim that mathematical principles are true. Presumably he thinks that further conditions must be satisfied (over and above verification) in order to show that principles are true. Let us ask what these further conditions might be and critically evaluate any suggestions that might be forthcoming at this point. In order to do this let us briefly review the theory of knowledge we have indicated above and to call attention to several points which it has not yet been necessary to discuss.

Our basic view is that a theory including its mathematical postulates is justified by the evidence providing that the observable consequences of the theory are verified. We have stressed above that the principles of the theory are justified by the verification process only if they actually are essential in deductions of the observable consequences which are verified. We now wish to make two further points. First, we do not regard logically inconsistent theories as justified by the evidence. If a theory is inconsistent that it is not possible for all of the statements it contains to be true. In such a case verification cannot confirm a theory since the theory will have consequences which conflict with observed data as well as consequences which agree with observed data. Second, we note that the expression "observable consequences of a theory" may be somewhat misleading. Strictly speaking it may not be possible to deduce from the statements of a theory any statements about observable data. Consider, for example, Darwin's theory that evolution of species proceeds in accordance with processes of natural and sexual selection. In order to make deductions concerning observable data from this theory, the theory would have to be supplemented by many statements concerning the particular conditions under which the evolution of particular species occurred. In brief, when we have spoken of observable consequences of a theory, what we have referred to are statements about observable data which are deducible from the statements of the theory when supplemented about other statements about particular conditions. Now, it should be noted that sometimes in deriving statements about observable data from theories, scientists sometimes supplement the theory with statements which are known to be false. Such statements are called simplifying assumptions. As an example, consider an explanation of the rate of motion of a ball on an inclined plane in which the observation statements are deduced with the aid of the assumption that the force of friction acting between the ball and the surface of the plane is nil. In light of the fact that simplifying assumptions are utilized in deriving observable consequences from theories we cannot say that all of the statements which enter into the deduction of observable consequences are confirmed. Indeed, many of the statements describing particular conditions are known independently and so are not confirmed when they are used in the process of confirming a theory. And the simplifying assumptions are not confirmed either. However, the theoretical assumptions including the mathematical principles are confirmed. Our view is that mathematical principles such as the axioms of real number theory are known to be true because

(1) so far as we know they are logically consistent,
(2) they have been strongly confirmed empirically and
(3) they would continue to be confirmed even if some of the

other substantive theoretical principles are rejected in favor of alternatives.

57. Let us now consider whether the above list of three items should be supplemented by addition of the following conditions: (i) theoretical principles which utilize impredicatively specified concepts cannot be known, (ii) theoretical principles which imply that sets exist cannot be known, (iii) theoretical principles which imply that an infinite number of objects exist cannot be known and (iv) theoretical principles which imply the existence of "non-constructive" entities cannot be known. In our view none of these additional conditions is warranted. We shall briefly indicate our reasons for taking this position.

The suggestion that specifications of objects in mathematics must be predicative has been widely discussed.[4] In our view, to rule out such definitions seems unwarranted. We see nothing objectionable in identifying the least upper bounds of certain sets of rational numberrs as real numbers. As noted earlier, Poincaré seems to have regarded such definitions as being guilty of some sort of ambiguity.[5] And Russell thought that such definitions violated the "vicious circle principle" which suggests that he thought that they were in some sense "circular definitions." But, as Quine has pointed out, impredicative definitions are not circular definitions. They do not smuggle the term to be defined implicitly into the definition.[6] Nor do such expressions as the least upper bound of a set of objects satisfying some condition C appear to be particularly ambiguous.

It has also been argued by Poincaré that avoidance of impredicative specifications of sets is justified as this is the correct way to avoid paradoxes such as Russell's paradox.[7] However, as many people have noted, the paradoxes can be avoided without making this restriction. There seems to be no sound argument which establishes that this is the "correct" way of avoiding the paradoxes. Restricting Frege's axiom of abstraction seems to be equally correct.

Conditions (ii) and (iii) above are similar in that they both rest on claims to the effect that statements of the sort that would violate these conditions are meaningless. Abraham Robinson said "I must regard a theory which refers to an infinite totality as *meaningless* in the sense that its terms and sentences cannot possess the direct interpretation in an actual structure that we should expect them to have by analogy with concrete (e.g., empirical) situations.[8] Nelson Goodman has claimed that references to classes are "incomprehensible."[9] But Robinson's claim seems simply to be mistaken. Each term in a mathematical theory, such as the theory of real numbers can, so far as I can see, "possess a direct interpretation." Terms like π and $\sqrt{2}$ can be interpreted as names of numbers. Sentences containing these terms can be interpreted as asserting that these numbers have certain properties or stand in certain relations, etc. Robinson's reference to actual structures and concrete situations suggests that he is saying that terms of theories which refer to infinite totalities cannot be construed as names of objects and, at the same time, correlated "directly" to sense perceptions. In other words, he may be saying that there are no sensory criteria which constitute sufficient conditions of something's being π or $\sqrt{2}$. But, if our arguments above are correct, then the same would be true even if our mathematical theories were restricted so as to imply only that there are a finite number of numbers. The point is that numbers are not directly observable. They cannot be identified with specific strokes along the lines suggested by Hilbert. The semantic criteria which Robinson may be suggesting rule out finite as well as infinite totalities so far as mathematical entities are concerned. Further, many other scientific terms would be ruled out also as we are not in a position to give sufficient conditions for the application of these terms which would satisfy extreme empiricist semantical rules. Consider, for example, biological terms such as "genotype" or "species."

However, if one is prepared to admit that theoretical terms such as "genotype" need only be "partially interpreted" by reference to sense perceptions in order to be meaningful then it appears that terms such as "continuity", "Cauchy sequence" $\sqrt{2}$ or π from theories which refer to infinite totalities can also be partially interpreted and so meaningful. For example, empirical criteria can be given for approximating to π or $\sqrt{2}$. In any case, Robinson's claim seems to be mistaken. If some mathematical terms are meaningful in a direct or literal sense then there seem to be no reasons for claiming that terms of theories which refer to infinite totalities are not meaningful also.

Nelson Goodman explains his objection more thoroughly. He conceives of the universe as consisting of individuals. Many objects of our experience are to be conceived as heaps or "sums" of individuals. The only collections of which Goodman can conceive are such sums. Set theory is incomprehensible to Goodman because it attributes existence to objects that are not mere sums of individuals in the following sense: Objects may be conceived as sums of individuals providing that no two objects consist entirely of the same individuals. Set theory violates this condition. For example, in set theory the object $\{a,b\}$ is not the same object as $\{\{a,b\}\}$ even though both objects are "built up" out of the same individuals, namely of a and b. The former object has two members namely a and b. The latter object has only one member namely, $\{a,b\}$. Since sets are identical if and only if they have the same members these two objects are not identical.

My response to Goodman is to say that this feature of set theory does not lead me to say that set theory is incomprehensible and to argue for this in terms of examples. A mathematics class in school, for example, may consist of exactly the same individuals as an english class. But, if someone said that the mathematics class is not the same as the english class this would not be incomprehensible. Or consider the following possibility, a and b above might be different populations and the set $\{a,b\}$ might be a species. On the other hand the set $\{\{a,b\}\}$ whose member is the set $\{a,b\}$ would be a genus. In the context of biological theory one might want to say of the genus that at the present time it has only one member, namely the species $\{a,b\}$ but that it is not identical with this species. Other examples where it makes sense to speak of different objects constituted of the same individuals could be given also. I do not, of course, hope to persuade Goodman by examples such as these to modify his position. He would, no doubt, try to provide translations of the statements that I used in giving these examples so as to show that where there are different objects there are also different individuals constituting them. He would thus hope to account for the meaningfulness of the examples I have given without having to allow that there are sets in the sense to which he objects. However, as we have been convinced of the existence of sets on other grounds (through the empirical confirmation of mathematical principles which imply that sets exist) we shall not try to refute him at this point by showing that possible translations which he might provide are unacceptable to us. Our point in giving these examples was merely to show that, to us at any rate, statements implying the existence of sets are quite comprehensible. They would remain so even should a translation along the lines that Goodman would accept prove unavailable. So far as I can see the sort of consideration that Goodman offers for finding reference to sets incomprehensible is unconvincing. It does not provide adequate grounds for adopting criterion (ii) as a criterion of knowledge.

With respect to criterion (iv), if "constructive" is to be understood as suggested by the intuitionists, then we reject this criterion because we do not believe that introspection is a particularly reliable way of getting knowledge. We do not wish to deny that introspection may have some value as a method of obtaining self-knowledge. But, even in this case, it should be supplemented by more "objective" procedures.

The strongest argument that can be given for adopting any of the criteria (i)–(iv) is perhaps that since the resulting mathematical theories would be weaker than mathematical theories which we presently claim to know, i.e., real number theory, parts of set theory, etc., such theories are even less likely to contain logical inconsistencies. But even this argument does not convince us to adopt these principles. Even the weakened theories could not be proved to be consistent and so it is not clear that the increased security we might gain from weakening our mathematical principles in accord with criteria (i)–(iv) is of sufficient degree to warrant the claim that we really do not know such principles as the axioms of real number theory, the axiom of choice, etc.

In considering principles (i) to (iv) we have been investigating suggestions that might be offered for strengthening the conditions that we say must be satisfied in order to have knowledge of mathematical principles (over and above the three conditions mentioned in section 56). We do not see any reason which is sufficient for strengthening the conditions for mathematical knowledge by adding any of these requirements. Of course we do not claim to have considered all possible ways of strengthening the conditions for knowledge. But these four additional conditions were suggested to us by recent discussion in the philosophy of mathematics. Since we reject the idea that the conditions for knowledge should be strengthened with respect to any of these principles then we feel justified in saying with respect to the criticism discussed in section 56 that the claim that we are merely justified in using mathematical principles but not in believing that they are true is unwarranted.

In similar manner we might adduce further arguments against skepticism. The skeptic had adopted conditions for knowledge which, we believe, are unjustifiable. What these conditions are varies with the skeptic. Some skeptics have apparently taken the position that if a person knows that P then he must be certain that P either because P is necessarily true or because P has been certified by some infallible method. Then the realization that there are no infallible methods of certification and no necessary truths leads to skepticism. But, in our view, this argument for skepticism is undermined when we refuse to allow that in order to have knowledge that P one must be certain that P. Other skeptics with regard to mathematical knowledge have adopted conditions for knowledge which are, in our judgement, too extremely empiricist. They have held that one cannot have knowledge of some object unless that object is perceived by the senses. We do not believe that there is any warrant for this assumption. But, without some assumption such as this there seems to be no justification for claim that mathematical objects are all imaginary. Some skeptics have been led to their position through the mistaken acceptance of such conditions of knowledge.

Our case now rests. We believe that in this work we have shown that mathematical principles must be interpreted literally and that so interpreted they imply the existence of queer entities and that we have empirical knowledge of mathematical principles. Philosophers such as Vaihinger or Körner who have opted for skepticism are, we believe, mistaken. While the view that mathematical knowledge is empirical may strike some people as shocking, it should not be. We have already noted that our view has been strongly influenced by Quine and Putnam. And even Körner apparently would agree with us in our claim that mathematical principles are justified by an empirical procedure.[10] However, he would reject our claim that the justification of mathematical principles implies the existence of queer entities. His position thus seems quite close to Vaihinger. As we have indicated, we believe that skepticism with respect to the view that mathematical principles are true is unwarranted. Another philosopher who has taken a position quite close to that expressed here is Lazlo Kalmar. He said, "Why do we not confess that mathematics, like other sciences, is ultimately based upon and has to be tested in practice?"[11] And goes on to argue that we should include "inductive methods" in mathematics.

NOTES

1. Körner, (38) pp. 174–75.

2. Putnam, (64) p. 271.
3. Putnam, (68) pp. 54–5.
4. Chihara, (14).
5. See our prior discussion of impredicative definitions in section 29.
6. Quine, (73) p. 242.
7. Poincaré, (61) p. 177 f.
8. Robinson, (78) p. 231.
9. Goodman, (23) p. 198.
10. Körner, (39).
11. Kalmar, (37).

References

1 Ayer, Alfred Jules, *Language Truth and Logic* (New York, Dover Publications, Inc., 1946).

2 Bar-Hillel, Yehoshua, *Logic, Methodology and Philosophy of Science* (Amsterdam, North-Holland Publishing Co., 1965).

3 Benacerraf, Paul and Putnam, Hilary, *Philosophy of Mathematics, Selected Readings* (Englewood Cliffs, Prentice-Hall, Inc., 1964).

4 Brandt, Richard B. and Nagel, Ernest, *Meaning and Knowledge, Systematic Readings in Epistemology* (New York, Harcourt, Brace and World, 1965).

5 Brouwer, L.E.J., "Intuitionism and Formalism", *Philosophy of Mathematics Selected Readings*, ed. Benacerraf, Paul and Putnam, Hilary (Englewood Cliffs, Prentice-Hall, Inc., 1964), pp. 66–77. This is the inaugural address at the University of Amsterdam, read October 14, 1912.

6 ——, "Consciousness, Philosophy, and Mathematics", *Philosophy of Mathematics, Selected Readings*, ed. Benacerraf, Paul and Putnam, Hilary (Englewood Cliffs, Prentice-Hall, Inc., 1964), pp. 78–84.

7 Brown, Stephen I., "Mathematics and Humanistic Themes: Sum Considerations", *Educational Theory*, Vol. 23, No. 3, Summer 1973.

8 Buchler, Justus, *Philosophical Writings of Peirce* (New York, Dover Books, 1955).

9 Carnap, Rudolf, "The Logicist Foundations of Mathematics", *Philosophy of Mathematics, Selected Readings*, ed. Benacerraf, Paul and Putnam, Hilary (Englewood Cliffs, Prentice-Hall, Inc., 1964), pp. 31–41. This essay formed part of a symposium on the foundations of mathematics which appears in Erkenntnis (1931).

10 ——, *Foundations of Logic and Mathematics*, Volume 1, Number 3, International Encyclopedia of Unified Science (Chicago, University of Chicago Press, 1939).

11 ——, *Meaning and Necessity, A Study in Semantics and Modal Logic* (Chicago, University of Chicago Press, 1949).

12 ——, *The Logical Syntax of Language* (Paterson, Littlefield, Adams and Co., 1959).

13 ——, "Empiricism, Semantics and Ontology," *Philosophy of Mathematics Selected Readings* (Englewood Cliffs, Prentice-Hall, Inc., 1964). Reprinted from Rudolf Carnap, *Meaning and Necessity* 2nd ed. (Chicago: The University of Chicago Press, 1956) and from *Revue Internationale de Philosophie*, No. 4 (1950), pp. 20–40.

14 Chihara, Charles S., *Ontology and the Vicious Circle Principle* (Ithaca, Cornell University Press, 1973).

15 Copi, Irving M., *The Theory of Logical Types* (London, Routledge and Kegan Paul, 1971).

16 Curry, Haskell, *Outlines of a Formalist Philosophy of Mathematics* (Amsterdam, North-Holland Publishing Company, 1951).

17 ——, "Remarks on the Definition and Nature of Mathematics", *Philosophy of Mathematics, Selected Readings*, Benacerraf, Paul and Putnam, Hilary (Englewood Cliffs, Prentice Hall, Inc., 1964).

18 Fraenkel, Abraham, and Bar-Hillel, Yehoshua, and Levy, Azriel and Van Dalen, Dirk, *Foundations of Set Theory* (Amsterdam, North-Holland Publishing Company, 1973).

19 Frege, Gottlob, *The Foundations of Arithmetic* (New York, Philosophical Library, 1953).

20 Geach, Peter and Black, Max, *Philosophical Writings of Gottlob Frege* (Oxford, Basil Blackwell, 1952).

21 Gödel, Kurt, "Russell's Mathematical Logic", *Philosophy of Mathematics, Selected Readings*, Benacerraf, Paul and Putnam, Hilary (Englewood Cliffs, Prentice-Hall, Inc., 1964). Reprinted from Paul A. Schilpp, ed., *The Philosophy of Bertrand Russell*, The Library of Living Philosophers.

22 ——, "What is Cantor's Continuum Problem", *Philosophy of Mathematics, Selected Readings*, Benacerraf, Paul and Putnam, Hilary (Englewood Cliffs, Prentice-Hall, Inc., 1964). This is a revised and expanded version of a paper of the same title which appears in *The American Mathematical Monthly*, 54 (1947), 515–25.

23 Goodman, Nelson, "A World of Individuals", *Philosophy of Mathematics, Selected Readings*, Benacerraf, Paul and Putnam, Hilary (Englewood Cliffs, Prentice-Hall, Inc., 1964).

24 Gottlieb, Dale, "A Method for Ontology, with Application to Numbers and Events," *The Journal of Philosophy*, Volume LXXIII, No. 18, October 21, 1976.

25 Hahn, Hans, "Conventionalism", *Logic and Philosophy, Selected Readings*, Iseminger, Gary (New York, Appleton-Century Crofts, 1968).

26 Hatcher, William S., *Foundations of Mathematics* (Philadelphia, W.B. Saunders Company, 1968).

27 Hempel, Carl, "On the Nature of Mathematical Truth", *Philosophy of Mathematics, Selected Readings*, Benacerraf, Paul and Putnam, Hilary (Englewood Cliffs, Prentice-Hall, Inc., 1964). Reprinted from *The American Mathematical Monthly*, Vol. 52 (1945).

28 ——, "Geometry and Empirical Science", *American Mathematical Monthly*, Vol. 52, pp. 7–17, 1945.

29 Heyting, Arend, "The Intuitionist Foundations of Mathematics", *Philosophy of Mathematics, Selected Readings*, Benacerraf, Paul and Putnam, Hilary (Englewood Cliffs, Prentice-Hall, Inc., 1964). Reprinted from symposium on the foundations of mathematics which appeared in Erkenntnis (1931).

30 ——, "Disputation", *Philosophy of Mathematics, Selected Readings*, Benacerraf, Paul and Putnam, Hilary (Englewood Cliffs, Prentice-Hall, Inc., 1964). Excerpted from Arend Heyting, *Intuitionism, An Introduction.*

31 ——, *Intuitionism, An Introduction* (Amsterdam, North-Holland Publishing Co., 1956).

32 ——, "Intuitionistic Views on the Nature of Mathematics", *Synthese*, Volume 27, Nos. 1/2, May/June 1974.

33 Hilbert, David, "On the Infinite", *Philosophy of Mathematics, Selected Readings*, Benacerraf, Paul and Putnam, Hilary (Englewood Cliffs, Prentice-Hall, Inc., 1964).

34 Iseminger, Gary, *Logic and Philosophy, Selected Readings* (New York, Appleton-Century-Crofts, 1968).

35 James, William, *Essays in Pragmatism* (New York, Hafner Publishing Co., 1948).

36 Kalmar, Laszlo, "Foundations of Mathematics—Whither Now?" *Problems in the Philosophy of Mathematics"*, Lakatos, Imre (Amsterdam, North-Holland Publishing Company, 1967).

37 Kitcher, Philip, "Hilbert's Epistemology", *Philosophy of Science*, Vol. 43, No. 1, March 1976.

38 Körner, Stephan, *The Philosophy of Mathematics* (New York, Harper Torchbooks, 1960).

39 ——, "An Empiricist Justification of Mathematics", *Logic, Methodology and Philosophy of Science*, Bar-Hillel, Yehoshua (Amsterdam, North-Holland Publishing Company, 1965).

40 Lakatos, Imre, "Proofs and Refutations", *The British Journal for the Philosophy of Science*, Vol. XIV, No. 53, 1963.

41 ——, "Infinite Regress and Foundations of Mathematics", *Aristotelian Society*, Supplementary Volume, XXXVI, 1962.

42 ——, "A Renaissance of Empiricism in the Recent Philosophy of Mathematics", *The British Journal for the Philosophy of Science*, Vol. 27, No. 3, September 1976.

43 Leach, James, and Butts, Robert, and Pearce, Glenn, *Science, Decision and Value* (Dordrecht, Reidel Publishing Co., 1972).

44 Lehman, H., "Queer Arithmetics", *Australasian Journal of Philosophy*, Vol. 48, No. 1, May 1970.

45 Leibniz, G.W., "Letter to Canon Foucher", *Leibniz Selections*, Weiner, Philip P. (New York, Charles Scribners and Sons, 1951).

46 Levy, Stephen H., "On the Nature of Arithmetic Truths", paper read at meetings of the American Philosophical Association, 1971.

47 Lewis, Clarence Irving, "A Pragmatic Conception of the A Priori", *Meaning and Knowledge, Systematic Readings in Epistemology*, Brandt, R.B. and Nagel, E. (New York, Harcourt, Brace and World, 1965).

48 Locke, John, *An Essay Concerning Human Understanding* (New York, Dover Publications, 1959).

49 Lyusternik, L.A., *Convex Figures and Polyhedra* (New York, Dover Publications, 1963).

50 Mackie, J.L., "Proof", *Proceedings of the Aristotelian Society*, Supplementary Volume XL, 1966.

51 Marsh, Robert C., *Bertrand Russell, Logic and Knowledge* (London, George Allen and Unwin Limited, 1956).

52 Massey, Gerald, "Tom, Dick and Harry, and All the King's Men", *American Philosophical Quarterly*, Volume 13, No. 2, April, 1976.

53 McCall, Storrs, "A Non-classical Theory of Truth, With an Appli-

cation to Intuitionism", *American Philosophical Quarterly*, Volume 7, No. 1, Jan. 1970.

54 Michalos, Alex, "Cost Benefit versus Expected Utility Acceptance Rules", *Science, Decision and Value*, Leach, James, and Butts, Robert and Pearce, Glenn (Dordrecht, Reidel Publishing Co., 1972).

55 Mill, John S., *John Stuart Mill's Philosophy of Scientific Method*, Nagel, E. (New York, Hafner Publishing Company, 1950).

56 Moise, E., *Elementary Geometry from an Advanced Standpoint* (Reading, Addison-Wesley Publishing Co., Inc., 1963).

57 Nagel, Ernest, "Logic Without Ontology", *Philosophy of Mathematics, Selected Readings*, Benacerraf, Paul, and Putnam, Hilary (Englewood Cliffs, Prentice-Hall, Inc., 1964).

58 Nelson, J.O., "Is Material Implication Inferentially Harmless", *Mind*, Vol. LXXV, No. 300, October, 1966.

59 Peirce, Charles Sanders, "The Fixation of Belief", *Philosophical Writings of Peirce*, Buchler, Justus (New York, Dover Books, 1955).

60 Poincaré, Henri, *Science and Hypothesis* (New York, Dover Publications, 1952).

61 Poincaré, Henri, *Science and Method* (New York, Dover Publications, 1952).

62 ——, *Mathematics and Science; Last Essays* (New York, Dover Publications, Inc. 1963).

63 Popper, Karl, *Objective Knowledge, An Evolutionary Approach* (Oxford, Oxford University Press, 1972).

64 Putnam, Hilary, "Degree of Confirmation and Inductive Logic", *Mathematics, Matter and Method, Philosophical Papers*, Volume 1 (Cambridge, Cambridge University Press, 1975).

65 ——, "Truth and Necessity in Mathematics", *Mathematics, Matter and Method, Philosophical Papers*, Volume 1 (Cambridge, Cambridge University Press, 1975).

66 ——, "Logic of Quantum Mechanics", *Mathematics, Matter and Method, Philosophical Papers*, Volume 1 (Cambridge, Cambridge University Press, 1975).

67 ——, "The Thesis That Mathematics is Logic", *Mathematics, Matter and Method, Philosophical Papers*, Volume 1 (Cambridge, Cambridge University Press, 1975).

68 ——, *Philosophy of Logic* (New York, Harper Torchbooks, 1971).

69 Quine, Willard Van Orman, "On What There Is", *From a Logical Point of View* (Cambridge, Harvard University Press, 1953).

70 ——, "Two Dogmas of Empiricism", *From a Logical Point of View* (Cambridge, Harvard University Press, 1953.)

71 ——, "Truth by Convention", *Philosophy of Mathematics, Selected Readings*, Benacerraf, Paul and Putnam, Hilary (Englewood Cliffs, Prentice-Hall Inc., 1964).

72 ——, *The Ways of Paradox and Other Essays* (New York, Random House, 1966).

73 ——, *Set Theory and Its Logic*, Revised Edition (Cambridge, Harvard University Press, 1969).

74 ——, *Philosophy of Logic* (Englewood Cliffs, Prentice Hall, 1970).

75 Ramsey, Frank Plumpton, "The Foundations of Mathematics", *The*

Foundations of Mathematics and Other Logical Essays (London, Routledge and Kegan Paul, 1931).

76 Rescher, Nicholas, *The Coherence Theory of Truth* (Oxford, Clarenden Press, 1973).

77 Resnik, Michael D., "On the Philosophical Significance of Consistency Proofs", *Journal of Philosophical Logic* 3 (1974, pp. 133–147.)

78 Robinson, Abraham, "Formalism 64", *Logic, Methodology and Philosophy of Science*", Bar-Hillel, Yehoshua (Amsterdam, North-Holland Publishing Company, 1965).

79 ——, *Non-Standard Analysis* (Amsterdam, North-Holland Publishing Company, 1966.

80 Rudin, Walter, *Principles of Mathematical Analysis* (New York, McGraw Hill Book Company, 1964).

81 Russell, Bertrand, "Mathematical Logic as Based on The Theory of Types", in *Bertrand Russell, Logic and Knowledge*, Marsh, Robert C. (London, George Allen and Unwin Ltd., 1956).

82 ——, *Mysticism and Logic (Garden City, Doubleday Anchor Books).*

83 ——, *Introduction to Mathematical Philosophy* (London, George Allen and Unwin Ltd., 1919).

84 ——, *The Principles of Mathematics*, second edition (London, George Allen and Unwin Ltd., 1937).

85 Scheffler, Israel, *The Anatomy of Inquiry* (London, Routledge and Kegan Paul Ltd., 1964).

86 Simmons, George F., *Topology and Modern Analysis* (New York, McGraw-Hill Book Company Incorporated, 1963).

87 Steiner, Mark, *Mathematical Knowledge* (Ithaca, Cornell University Press, 1975).

88 Troelstra, A.S., *Principles of Intuitionism* (New York, Springer-Verlag, 1969).

89 Vaihinger, Hans, *Philosophy of the As-If* (London, Routledge and Kegan Paul, 1935).

90 Waismann, Friedrich, *Introduction to Mathematical Thinking* (New York, Harper and Brothers, 1959).

91 Wittgenstein, Ludwig, *Remarks on the Foundations of Mathematics* (Cambridge, The M.I.T. Press, 1956).

NAME INDEX

SUBJECT INDEX